SUSTAINABLE EXTRACTIVE SECTOR MANAGEMENT

ISSUES AND PROSPECTS

ZIK IGBADI BONIWE

ABOUT THE AUTHOR

Mr. Zik Igbadi Boniwe was born to the family of late Mr. Thaddeus Nwagboniwe Igbadi and Princess Regina Nwanefuluno Igbadi (Nee Ezechie) of Ogbeuchi Quarters, Ewulu Kingdom in Aniocha South Local Government Area of Delta State, in the oil-producing Niger Delta region of Nigeria.

He was raised by his grandfather, the late His Royal Highness Obi John Ezechie 1 (MON), the Obi (King) of Ewulu Kingdom in Aniocha South Local Government Area of Delta State, in the Niger Delta region of Nigeria. This is where he learnt how to resolve community-related issues, an experience that shaped his future engagement in resolving seemingly intractable issues.

He holds a Bachelor of Arts degree in Liberal Arts from the University of the State of New York; a Master of Arts Degree in Organizational Management from the University of Phoenix, Arizona; and a Master of Law in International Business Transactions in Natural Resources Law and Policy from the Centre for Energy, Petroleum and Mineral Laws and Policy, Faculty of Law and Accountancy, the University of Dundee, Scotland.

Mr. Boniwe is an Industrial/Management Consultant. He served as the Executive Managing Director of MAZIK Resources International Limited. He also served as the Programs Director of African Network for Environment and Economic Justice (ANEEJ), where he collaborated with some oil-producing host communities in the Niger Delta to ensure enactment of laws on the management of 13 per cent oil revenue derivation flowing from the Federation Account to oil-producing states for the development of host and impacted communities.

Mr. Boniwe's other Extractive-Industry-related publications are:

1. Towards Sustainable Development in the Solid Minerals, Oil, Gas and Energy Sectors (2004).
2. Earth's Natural Resources: A New Paradigm for Sustainable Development (2007).

ABOUT THE BOOK

The planet earth is fortified with abundant natural resources such as land and its contents; air with its constituent elements and water with both living and non-living things. These natural resources create a system of ecosystem regeneration for sustainability. Through history, humanity has depended on these natural resources for sustenance.

Extractive companies involved in harnessing these natural resources for the benefit of humanity have advanced technological adeptness that has steered towards massive exploration, exploitation, processing, usage and disposal of these resources. The associated activities of natural resources development have improved and also negatively impacted quality of life from prehistoric age to modern industrial society.

No doubt, the extractive sector has positively contributed to technological advancement, improved education and incomes, access to health care which hold ever-greater promise for longer, healthier, more secure lives both in developed and some developing counties. However, there is also a widespread sense of instability in the world today from the activities of the extractive companies in livelihoods, in personal security, in the environment and in global politics in almost all the developing countries where the extractive industries operate.

Sustainable Extractive Sector Management: Issues and Prospects delved into both the positive and negative impacts of the extractive sector on the governments, the extractive companies and the hosts and impacted communities by taking a comprehensive look at the conflicts that encumber sustainable extractive sector management and enunciates the critical issues that need to be addressed or reversed with implementation strategies. This will avert continuous disruptions of extractive industries operations and improve quality of lives of all stakeholders to ensure sustainable socio-economic development through mutual collaboration of key stakeholder groups.

AuthorHouse™
1663 Liberty Drive
Bloomington, IN 47403
www.authorhouse.com
Phone: 1 (800) 839-8640

Published by AuthorHouse 01/11/2018

ISBN: 978-1-5462-2401-3 (sc)
ISBN: 978-1-5462-2402-0 (e)

Print information available on the last page.

Any people depicted in stock imagery provided by Thinkstock are models,
and such images are being used for illustrative purposes only.
Certain stock imagery © Thinkstock.

This book is printed on acid-free paper.

Because of the dynamic nature of the Internet, any web addresses or links contained in this book may have changed
since publication and may no longer be valid. The views expressed in this work are solely those of the author and do not
necessarily reflect the views of the publisher, and the publisher hereby disclaims any responsibility for them.

authorHOUSE®

DEDICATION

This book, *"Sustainable Extractive Sector Management: Issues and Prospects"* is dedicated to every person who has participated in the development of the extractive sector; and those who live in and around the host and impacted communities where extractive industries operate that has been negatively affected by the activities of the extractive companies and to those who appreciate the plights of the victims of the extractive industries and have joined efforts to work towards the achievement of sustainable socio-economic development that ameliorates those plights.

PREFACE

The extractive sector in most developing countries has continued to be characterized by persistent frictions among key stakeholders with non-existent access to social security coupled with insufficient employment and employment benefits, inequalities in outcomes for women, youths and specific marginalised groups of people.

Overall, the extractive sector has failed to help majority of individuals and their families living in the host and impacted communities in their bid to escape poverty; but in most cases has planted misery and cognate anguish.

As a result of these reasons, there is the urgent need for governments and other critical stakeholders to develop strategic policies and programmes aimed at mitigating the critical issues and challenges of human rights abuses, environmental pollution and degradation, wrongful deaths, intra and inter communal conflicts, poverty as a result of environmental despoliation, conversion of community residential and farm lands without payment of adequate and timely compensation, and also global warming occasioned by the activities of the extractive companies.

However, in order to formulate and implement the well-deserved effective interventions in the extractive sector management, it is crucial to understand the nature of these challenges besetting the sector and how to match them with appropriate policies and institutional responses.

Though a lot has been written about the extractive sector in general and on individual extractive industries in particular, however, the majority of the academic literature in this field tend to be too technical for most policy-makers to access given their limited time and competing demands. This is the reason, *Sustainable Extractive Sector Management: Issues and Prospects"* is both timely and highly relevant to the needs of governments and other key stakeholders. This volume has been put together to provide a robust insight into key areas that are relevant to be explored in order to come up with more effective strategies to ensure mutually beneficial extractive sector development.

This book, *"Sustainable Extractive Sector Management: Issues and Prospects"*, which is written at this critical moment in time that could be described as cross-road point in extractive industries (EI) development especially in developing countries of Africa, Asia, Latin America and the Oceania, is intended to bring to the fore the gains to humanity from the extractive sector and a litany of bitter impacts of EI development, resulting from extraordinary exclusively profit-oriented legal and regulatory frameworks designed and probably driven by age-long colonial colouration and their negative impacts on the host jurisdictions.

This book is written as the nexus and beacon of hope through which the fundamental issues connected to unsustainable natural resources development that thus far represents a frightening trend to both the host and impacted communities in particular and host developing nations at large have been x-rayed with the aim to reverse the seemingly intractable negative consequences of extractive industries operations on mostly the poor communities in developing countries and guarantee sustainable natural resources development.

With this in mind, the book endeavours to point the key stakeholders towards some fundamental issues that must be addressed in order to stop and reverse the agony visibly evident in the extractive industries host and impacted communities, through the economic activities of the extractive companies operating in developing countries.

This is achieved by looking at some of the past externally driven profit-oriented regulatory frameworks and their negative social, economic and environmental implications that underscore the need to introduce more pragmatic legal, fiscal and regulatory frameworks that will more appropriately reinforce the institutional and financial capacities of host countries to ensure effective contract negotiation, project implementation, monitoring, evaluation and enforcement of agreements with multinational companies and Bilateral Investment Treaties (BITs) with their home states, thereby being well equipped to introduce necessary corrective actions that will reduce rancorous and conflict-laden relationships and be at the driver-seat to manage their natural resources and improve cooperation among stakeholders in this economic sector.

To reverse the apparently "no-friend-lost" clandestine relationship currently existing in some extractive jurisdictions will require both direct and active participation of key stakeholders in decisions regarding the management of the extractive sector. However, it is important that the stakeholders who would be participating in decisions regarding the management of the extractive sector have a very good knowledge of the extractive sector. This in addition to promoting increased awareness and understanding of extractive industries-related issues, will promote learning, and trigger appropriate actions among the stakeholders. This is an example of good management practice that is appropriate to be adapted for implementation in the extractive industries in a variety of local, culturally and environmental sensitivity areas.

This book is divided into eight Chapters. Chapter one is the introductory chapter which sets forth the groundwork for the issues discussed in the book. Chapter two introduces the stakeholders in the extractive sector, which include the government, the bilateral and multilateral partners, the extractive companies, the host communities, the civil society, the media, and the employees (Unions). This chapter also briefly discusses how each of these major actors in the extractive sector interact, influence or is influenced by the sector.

In chapter three, the book looks at selected extractive jurisdictions in developing countries and x-rays how the extractive industries have benefitted the people and the anguish suffered by the residents of the host and impacted communities. In chapter four, beginning from their foundation, the activities of the Bretton Woods institutions, the World Bank and its component agencies and the International Monetary Fund (IMF) as they impact the extractive industries are brought to the fore. Here the book reviews how the activities of these institutions have negatively impacted host governments and communities.

Chapter five discusses some attempts towards achieving sustainable development in the extractive sector. Chapter six discusses the impediments to sustainable development of the extractive sector. Here again, some very influential governments and institutions that have not played leading roles or have not lived up to their responsibilities are reminded to fulfil such responsibilities.

Chapter seven presents the sustainable actions and activities required for the extractive industries to benefit all key stakeholders and chapter eight deals with the implementation of such actions.

It is my hope that this book is welcomed by the international community, development partners, policy-makers, NGOs, trade unions, industrialists, consultants, employees' organizations, host communities, the media, teachers and students, involved in helping to provide mutually sustaining beneficial strategic approach to mitigating the most fundamental issues facing extractive sector development across developing countries the world over.

ZIK IGBADI BONIWE
JANUARY, 2018

FORWARD

Successive books have been written on the natural resources sector development with seemly enthusiastic praises of the sector for its positive economic, social and technological contributions to human development and conversely for the unrelenting indictment of the extractive companies for their negative human rights and environmental impacts especially in the developing countries of the world where they operate.

This particular book "***Sustainable Extractive Sector Management: Issues and Prospects"*** has delved into both the positive and negative impacts of the extractive sector on the governments, the extractive companies and the hosts and impacted communities. The author ably documented in precise term that the extractive companies have contributed immensely to societal development with most people in most developed countries doing steadily better in human development, but, the same could not be proffered for the developing countries.

No doubt, the extractive sector has positively contributed to technological advancement, improved education and incomes, access to health care which hold ever-greater promise for longer, healthier, more secure lives both in developed and some developing counties. However, there is also a widespread sense of instability in the world today from the activities of the extractive companies in livelihoods, in personal security, in the environment and in global politics in almost all the developing countries where the extractive industries operate.

In its eight chapters, this book has been able to enunciate critical issues that traverse the extractive sector and the need to critically address these issues by the key industry stakeholders to reverse the increasing helplessness of the affected communities. The need to prevent continuous disruptions of extractive industries operations has also been advanced by the author in this book, as a way to ensure sustainable socio-economic development through the contributions of the extractive sector.

Mr Boniwe in this book informs that People's well-being is influenced greatly by the larger freedoms within which they live and by their ability to respond to and recover from adverse events—natural or human-made.

Further, the book examined how the activities of the extractive companies have left people socially, physically and psychologically impoverished; encouraging unresponsive state institutions to the plights of those in need of assistance, creating community threats and tensions, violent conflicts, neglect of public health, environmental damages, crime and discrimination, all accumulating to individual and community susceptibility.

According to Mr Boniwe, Natural and human-made disasters emanating from extractive companies' activities are inevitable, but proactive efforts must be made to mitigate their effects and to accelerate recovery that extends beyond immediate threats and shocks to address underlying causes and longer term impacts. He further indicates that global public goods and universal social goods that would correct or complement for more inclusive and sustainable minimum levels of social protection and commitments to the provision of social services by government and extractive sector operators are important public goods that can be included in the sustainable development goals of host and impacted communities to enhance their abilities to cope with adverse impacts of the extractive sector operations.

The book points out that the contributions of the extractive sector on critical aspects of human development as education, health and nutrition, and employment have been quickly undermined by environmental degradation, human rights abuses, insecurity of the people, natural disaster and decline in economic wellbeing of the hosts and impacted communities which are linked to the activities of the extractive companies.

The book furthermore enunciates that women bear the brunt of personal insecurity in developing countries extractive jurisdictions. Violence in the extractive jurisdictions infringes on the rights of women, and the feelings of personal insecurity restrict their ability to expand their freedoms, human security and sustainable livelihoods. The author prescribed the supporting measures to bring about needed changes in the activities of the extractive industries that will reverse women vulnerability and improve gender equality which is presented in this book as a cross-cutting issue in the extractive sector development.

Mr Boniwe recommended through this book that altering public negative perceptions of the extractive sector requires openness, transparency and accountability to the host and impacted communities, the poor and excluded, as well as the promotion of a positive dynamism between governance institutions and civic participation. He also proposed and identified the strategies to remove the barriers that hold people back from participating in development decisions.

This book takes a broader approach, emphasizing the close links between business and human rights in line with the UN Guiding Principles on Business and Human Rights; reducing Defenselessness and advancing sustainable development of human society especially the developing countries through the extractive sector.

Mr Boniwe introduces the concept of *Stakeholders Partnership Arrangement (SPA)* to describe the prospects of reversing vulnerability and erosion of human rights and economic growth, drawing attention to the risks of future decline in individual, community and national circumstances and achievements, and putting forward the strategic approach to implementing the SPA framework to ensure against threats and make human development progress more robust.

In presenting the SPA framework, Mr Boniwe acknowledges that the road to putting into practice the SPA approach will not be an easy thoroughfare. According to him, the road will appear to be long, rough and tenuous; however, it will end as a journey of renewed hope, excitement, and opportunities.

I, in all honesty, commend the efforts of Mr Zik Igbadi Boniwe in carrying out his in-depth research and documentation with high level of intellectual exposition in what could rightly be adjudged a pioneering effort through this book which is a must-read for extractive industries practitioners, government policy-makers, communities, students and lecturers of tertiary institutions, the media, civil society organizations, international financial institutions and the international community as a whole.

Dr Emmanuel O. Emmanuel.

ACKNOWLEDGEMENT

First, I thank Almighty God whose guidance has made this book possible. I also thank the staff, students and lecturers of University of Dundee, Centre for Energy, Petroleum and Mineral Law and Policy, who shaped my understanding of the extractive sector through the knowledge obtained from their orientation, support and lectures, while a student of the institution. To late Professor Thomas Wälde (1949–2008), who personally directed my course of study while a student at University of Dundee, you were a quintessential Lawyer and Professor beyond conventional thought. I thank you posthumously for the wealth of knowledge of the extractive sector bequeathed on me.

To a Senior Lecturer in Oil and Gas Law at University of Lagos, Dr Dayo Ayoade, I say thank you for your time and dedicated effort in reviewing this book and making your insightful recommendations and inputs for its improvement.

Words are not enough to appreciate the tremendous contributions of Dr Austin Onuoha, the Executive Director of African Centre for Corporate Responsibilities who not only did a wonderful review of the book but made concrete suggestions for its improvement. I thank you very much for your encouragement and support.

To my very good friend, Barrister Agatha Osieke, my appreciation knows no boundaries. Thank you!

I thank my senior brother Dr, Barrister Nat A. Igbadi (KSC), who made all this work come to fruition by delaying the pursuit of his educational career and working to sponsor my education from secondary school to the university level and making sure that I attained appreciable level of education before he could go back to school to face his own educational pursuit; who also made fabulous contribution by editing this book, making sure, as much as possible, that all 'I's are dotted and all 'T's crossed. God will bless everything related to you, AMEN!

I have had the privilege to be involved for over 16 years in research in the extractive sector focusing especially on the socio-economic, environmental and human rights angle of the extractive sector, the products of which have been applied in working to build the capacity of Civil Society Organisations and local communities in Nigeria on participatory decisions and management of the extractive sector for the benefit of all citizens. This was given impetus by my professional relationship with Reverend David Ugolor, and Comrade Leo Atakpu, the Executive Director and Deputy Executive Director of African Network for Environment and Economic Justice (ANEEJ), who provided me the opportunity to serve as the Programmes Director of ANEEJ, directly engaging with the host oil-producing communities in the Niger Delta and witness first-hand their plights; and also working with state legislative branch of government in some of the oil-producing states in the Niger Delta to ensure the drafting and passage of bill into law for the management of 13 per cent oil revenue derivation coming to the states government from the Federation Account, for the development of oil-producing and impacted communities. This opportunity also expanded to a consultancy to develop administrative and financial management system and designing activities and training manuals for Publish What You Pay (PWYP) Campaign in Nigeria.

I owe a great deal of gratitude to the Executive Director of Transparency and Economic Development Initiative (TEDI), Elder Emeka Ogazi, for engaging me as his technical consultant during his representation of African Civil Society Organisations in the World Bank Climate Investment Funds and also for developing his organisation's five-year strategic plan.

To the youths leadership of the 36 communities making up the Oil Prospecting License (OPL) 135 located in Delta North Senatorial District in Delta State Nigeria, I thank you for inviting me during your inauguration to share with you the concept of "Corporate Social Responsibility" and advance some strategies to positively mobilise yourself to speak with one voice and, build your capacity to effectively engage the oil prospecting company to avoid any rancorous relationship.

To individuals who work for extractive companies as Community Relations or Liaison Officers who have privately sort my assistance on the direction to effectively engage some agitated host communities youths and persuade them to douse their agitation, I thank you for reposing confidence in me to assist.

To everyone whose work was reviewed for this book, whether cited in the bibliography or not, I owe a great deal of appreciation. To this group, I sincerely apologise for any shortcomings that may have arisen from the usage of your intellectual property and accordingly, I assume full responsibility for any errors therein.

This book would not have seen the light of day without the commitment of the editor, publisher, proof-readers and printer.

The list of those that I owe a great deal of gratitude that shaped the outcome of this book is too long and can take a lot of the space dedicated to this book. Therefore, for those people who deserve to be mentioned but were not, please bear with me as I know that your contributions facilitated the successful publication of this book.

Finally, I thank my family, starting from my wife Isioma, my son, Kevin and my daughters Karen, Karisa and Kevinia. I say thank you for your immense contribution, allowing me time to concentrate and to completing this work. In addition, I must thank my mother, my brothers and sisters for their much love and unequivocal emotional support.

ACRONYMS

ACFOA	Australian Council for Overseas Aid
AfDB	African Development Bank
AFDL	Alliance of Democratic Forces for the Liberation of Congo-Zaire
ATCA	US Alien Tort Claims Act
BCDA	Basil Community Development Association
BCLO	Basil Community Liaison Officer
BIT	Bilateral Investment Treaty
BNPP	World Bank-Netherlands Partnership Program
CECD	Corporate Extractive Company Developers
CEOs	Chief Executive Officers
CMCA	Community Mine Continuation Agreement
CONAMA	Chilean National Environmental Commission
CSO	Civil Society Organisation
DIAMANG	*Companhia de Diamantes de Angola*
DNGM	Direction nationale de la géologieet des mines (in Mali)
DPs	Development partners
DRC	Democratic Republic of Congo
EAP	*Energia, Ambiente, y Poblacion*(Energy, Environment, and Population)
EDS	Economic Development Services
EFCA	Edo State Farmers' Cooperative Agency
EFCC	Economic and Financial Crimes Commission
EIA	Environmental Impact Assessment
EITI	Extractive Industry Transparency Initiative

FAO	Food and Agriculture Organization
FDI	Foreign Direct Investment
FNLA	*Frente Nacional de Libertação de Angola*
GHG	*Greenhouse Gas*
GSA	General Services Agency
HQCF	High Quality Cassava Flour
IBRD	International Bank for Reconstruction and Development
I C J	International Court of Justice
ICSID	International Centre for the Settlement of Investment Disputes
IDA	International Development Association
IFC	International Finance Corporation
ILO	International Labour Organisation
IMF	International Monetary Fund
IRC	International Rescue Committee: American Humanitarian Organisation
JTF	Joint Task Force (comprising Nigerian Air Force, Army and Navy)
LAWYER	Land, Air, Water, You, Environment, Religion
LGA	Local Government Area
MDGs	Millennium Development Goals
MEND	Movement for the Emancipation of the Niger Delta
MIGA	Multilateral Investment Guarantee Agency
M-II	Management Institute Initiative
MNCs	Multinational Corporations
MOGE	Myanmar Oil and Gas Company
MOSOP	Movement for the Survival of Ogoni People
MoA	Memorandum of Agreement
MoU	Memorandum of Understanding
MPLA	*Movimento Popular de Libertação de Angola*
NAFTA	North Atlantic Free Trade Agreement
NATO	North Atlantic Treaty Organisation
ND	Niger Delta

NDPI	Niger Delta Partnership Initiative
NGO	Non-Governmental Organization
NLD	National League for Democracy party in Myanmar
NUPENG	National Union of Petroleum and Gas
OML	Oil Mining Lease
OPIC	Overseas Private Investment Corporation
OPL	Oil Prospecting License
OTDF	Ok Tedi Development Foundation
OTFRDP	Ok Tedi Fly River Development Program
OTMIAA	Ok Tedi Mine Impacted Area Association
OTML	Ok Tedi Mining Limited
P4P	Partners for Peace in the Niger Delta
PENGASAN	Petroleum and Natural Gas Senior Association of Nigeria
PGK	PNG Kina
PIND	Foundation for Partnership Initiatives in the Niger Delta
PNG	Papua New Guinea
PNGSDP	Papua New Guinea Sustainable Development Program
PNGSDPL	Papua New Guinea Sustainable Development Program Limited
PPPs	Public Private Partnerships
PTT	Petroleum Authority of Thailand
PWYP	Publish What You Pay
RICO	Racketeer Influenced and Corrupt Organization Act
SAP	Structural Adjustment Programmes
SPA	Stakeholders Partnership Arrangement
RUF	Revolutionary United Front of Sierra Leone
SLORC	State Law and Order Restoration Council
UN	United Nations
UNDP	United Nations Development Programme
UNGPBHR	United Nations Guiding Principles on Business and Human Rights
UNGC	UN Global Compact

UNITA	*União para la Independência Total de Angola*
US	United States
US$	United States Dollar
VDF	Village Development Fund
VPC	Village Planning Committees
WB	World Bank
WBG	World Bank Group
WIM	Women in Mining
WPPDTF	Western Province People's Dividend Trust Fund
WTO	World Trade Organization

CONTENTS

**Sustainable Extractive Sector Management:
Issues and Prospects**

**Sustainable Extractive Sector Management:
Issues and Prospects**

**Sustainable Extractive Sector Management:
Issues and Prospects**

**Sustainable Extractive Sector Management:
Issues and Prospects**

The author is being paddled in a canoe during his research work in the Niger Delta area of Nigeria

CHAPTER ONE

The Perspective

Mineral resources generate significant profits for the countries in which they are located and developed, as well as for the extractive companies that explore, exploit and develop these resources. The exploitation and production activities of the extractive companies provide revenues to government in the form of royalties, license fees and taxes. The export of the mineral resources also generates the foreign exchange required by the host country to finance imports and procure the infrastructure needed for economic growth and development. The extractive companies provide local employment opportunities in both the extractive and allied industrial sectors, leading to increased incomes, better living, health and educational conditions, among others. Both the host communities and communities near and beyond the extractive area find the new infrastructure set up by the extractive companies very beneficial, and this increases key stakeholders' interests in the extractive sector.

However, mineral wealth creation from the extractive sector also raises some fundamental issues.
One of the cardinal legacy problems evident in the extractive sector that encumber healthy relationships between the host communities and the extractive companies is the extractive companies' penchant to making promises to the host communities that they do not keep or fulfill.

Promises made by the extractive companies to the host communities must be kept, honored, respected and fulfilled in this way there will exist an atmosphere of amity rather than the rancorous relationships that abound in the extractive sector.

Another apparent problem is the sustainable management of the extractive sector. For instant, if the revenues generated from the extraction of the natural resources are not well managed, this can lead to sectorial imbalance and macro-economic instability that could create disruption in the socio-economic and political systems, through possible inducement of emigration and creation of other forms of stresses on other natural resources sectors such as water and scramble for farm lands or illegal logging and harvesting of forest woods and poaching of protected animals and consequent damage to the ecosystem.

These problems are currently present where extractive companies operate in developing countries and has been exacerbating for many years unabated because the institutions, the required technical capacity and goodwill necessary to deal with them are conspicuously absent. This also has presented the extractive companies with seemingly protracted complications in dealing with the concerns of the local host and impacted communities.

This notwithstanding, some progressively-thinking extractive companies have built in some strategies in their corporate social responsibility framework to overcome these challenges knowing that their projected profit level will be compromised if these problems are ignored. However, some extractive companies depend solely on the host governments to resolve the problems or out rightly decide to do nothing about the problems hoping that one day all the problems will dissipate into the thin air. These are some of the challenges facing the stakeholders especially in the developing countries.

To resolve these seemingly intractable problems require an integrated, well-coordinated and articulated comprehensive framework which all well knowing is effectively lacking in the extractive developing countries hosting these extractive companies and their development activities. Though it might not be possible to effectively resolve all the problems facing the key stakeholders in the extractive sector, it will contribute in no small measure to recognizing and clarifying some dire issues, facilitating some critical thoughts and generating debates that can encumber progressive degeneration of relationships as it is today, among the critical stakeholders.

To start this process therefore, just imagine a situation that as you prepare to go to sleep on Sunday night you hear a government announcement inviting the key stakeholders in the extractive sector to gather in the community town hall at 9:00 on Monday morning which is the following day for critical deliberation and decisions regarding the management of the natural resources within the community. Envisage also that when you wake up on that Monday morning you hear the same government announcement that at 9: 00 A.M., a group of people comprising representatives of different stakeholders in the extractive sector will meet to decide new ways to develop earth's natural resources to ensure maximum benefit for all.

Imagine further that at 9:00 A.M, representatives of government, extractive industries, the media, other private sectors, host communities, the CSOs, the academia, special interest groups, World Bank Group, International Monetary Fund, African Development Bank, donor countries and agencies actually gather to take tangible decisions on how natural resources could be developed with minimal collateral damages, and how the wealth generated from this sector could be effectively used to achieve sustained socio-economic development.

Could this approach in anyway resemble the vision of late President John F. Kennedy of sending Americans to visit and explore the moon? Yes, Cosmonaut Yuri Gagarin of the former Soviet Union was in space as far back as April 12, 1961 and the American John Glen orbited the earth in 1962. President J.F. Kennedy's vision was achieved and today there is a permanent base-station in space where former "cold war" enemies (Soviet Union, now Russia Republic and the United States of America) exchange scientific information/data and ideas for the advancement of humanity. Could this collaborative effort not be considered as one of the greatest achievement by mankind and between two former cold war actors that nearly plunged the world into third world war?

Again, if man can develop airplanes, nuclear powered submarines/ships, intercontinental ballistic missiles, and can burrow through the rocks for more than one kilometer underneath the earth surface to extract natural resources from their resting place, what stops him from solving the problems associated with the process of extraction of the natural resources, and institutionalize prudent systems of management of the revenues there from as is the case with Norway, Canada, Shetland, Alaska, and other jurisdictions in the developed world where natural resources are developed by the extractive industries?

The extractive companies on their own have what is required to change the rancorous relationships existing between them and their host and impacted communities to a normalized and mutually beneficial relationship.

If the extractive sector operators and the other key stakeholders can agree to work together to change from the present "survival of the fittest" mentality, the "greed syndrome" and the present "I-don't-care attitude because it is not my backyard" which is endemic in the extractive sector, to a genuine ***Stakeholders Partnership Arrangement*** (SPA) aimed at prudently managing the entire process of natural resources development-from ideological stage through prospecting for the natural resources to managing the revenues derived from them, then humanity would have achieved the same or surpass the success level registered in the above mentioned areas with the development of the extractive sector.

This without any prevarication would affirm that humanity has the proven capability to change the way the extractive sector is currently being managed in the developing countries.

The extractive sector[1] has the potential to generate significant wealth that will put permanent smiles of happiness on the faces of the poor residents of the host communities and host developing countries where minerals extractions take place. The extractive sector also has the potential to serve as vital mechanism for sustainable growth and poverty alleviation through the huge revenues generated from royalties, taxation, and exports and employment. For these reasons and more, the extractive sector therefore ought to be a catalyst for the positive transformation of life; driving of economic growth, creation of jobs and reduction of poverty in resource-rich developing countries. Too often, however, these opportunities do not materialize and instead the extractive industries deliver as much or more damage to the host and impacted communities than benefits.

In many developing countries of the world where natural resources[2] have been or are being extracted, there have been tales of woeful environmental despoliation and fatigue, bionetwork destabilization, and human rights abuses that consist of unlawful arrests and illegal incarcerations, forced-labor, forced-displacements or arbitrary relocation of residents, rapes and extra-judicial killings with manifest linkages to ethnic cleansing and genocides, meted out to local residents of these natural resources development jurisdictions.

The income from natural resources exploitations have been deployed to fuel and sustain conflicts; illegal arms trafficking by clandestine operatives supplying arms to militaristic parties; acquisition and usage of weapons of mass destruction by warring groups and money laundering, create mordant poverty, causing untimely death of innocent citizens. The truth of the situation is that the natural resources sector has over the years come to represent in the real sense of it, a terrifying nemesis to both the host and impacted communities in the developing countries extractive jurisdictions.

There is much evidence in existence in both the solid minerals and the hydrocarbon (oil and gas) subsectors to support this assertion. For instance, in the solid minerals sector, there is the story of Ok Tedi in the South West Province of Papua New Guinea where B.H.P. Billiton of Australia that mined gold and copper in that jurisdiction was sued in the Supreme Court of Melbourne, Australia by the residents living along Ok Tedi River as well as the Fly River for polluting their water.

Chile is another case in point of a developing country that has suffered from extensive environmental pollution from the activities of extractive international mining companies and where local residents and host communities

---

[1] Oil, Gas and Solid Minerals
[2] The natural resources discussed in this book does not include timber, water etc. and is limited to only solid minerals and hydrocarbons (oil and gas).

successfully challenged mining companies in the Chanaral and Huasco cases for the remediation of years of tailing dumping, air and water pollution.

In Irian Jaya, Indonesia, Freeport McMoRan Inc., American company mining silver, copper and gold that for years allegedly polluted the environments in the host mining and impacted communities was sued in the United States of America for environmental damage and human rights violations. Another case worth mentioning here is the Omai Gold mine case in Guyana against Cambior Inc.; which is a Canadian-based mining company that operated in Guyana.

The case of opening a uranium mine at the Australian Kakadu national forest, which is a designated World Heritage Park Area, cannot go unmentioned[3].

Moreover, the world is not oblivious but exceedingly cognisant of the carnage of war financed with financial resources from natural resources. The UNITA rebel group of Angola under the leadership of Jonas Savimbi allegedly sold hundreds of millions of dollars' worth of supposedly stolen diamonds into the legitimate diamond trade channels every year, with absolute impunity, to sustain its insurgence in Angola.

The Ex-President of Liberia, Charles Taylor, masterminded one of the most horrible wars on the African continent, paying for it with financial resources from mostly Sierra Leonean traded diamonds. In addition, Sierra Leone and the Democratic Republic of Congo (DRC) in their own capacity are also not left out in the roll call of victims of natural resources-based conflict financing.

The diamonds used to generate the revenues to fund the conflicts are generally known as "Blood Diamonds" or "Conflict Diamonds"[4]. These "blood or "conflict diamonds" made the civil wars or insurgencies in Africa significantly more atrocious with heavy collateral damage and lasted longer than could ever have been the case without revenues generated from extraction of natural resources.

In the oil and gas sector, there is abundant evidence of human rights violations and environmental pollution in the Niger Delta of Nigeria supposedly linked to the Royal Dutch/Shell, Chevron, AGIP, ExxonMobil, ELF, and Texaco among others. There is the human rights violations and environmental pollution in the case of the Huaorani people of the Amazon rainforest in Ecuador allegedly committed by Texaco Petroleum Company. There is also the allegation of human rights violations and environmental pollution through the combined efforts

---

[3] Specific activities and impacts of the extractive sectors in these countries are discussed below.
[4] Conflict diamonds, or 'blood diamonds', are diamonds used by rebel movements to buy weapons and fuel war.

of Total S.A. and the Myanmar Oil and Gas Enterprises of Burma (Myanmar) in Myanmar. In addition, there is the alleged human rights violation and environmental pollution in the Cano-Limon oil field in the Arauca Department blamed on the consortium of Occidental Petroleum, Royal/Dutch Shell, and ECOPETROL and in the Cusiana-Cupiagua oil fields in the Casanare Department allegedly caused by the consortium of British Petroleum, ECOPETROL, Total and Triton in Colombia, among others. All these cases of environmental abuse and human rights violations mentioned above include extra-judicial executions, massacre, rapes, beatings, arbitrary arrests and unlawful detentions ostensibly linked to the extractive sector.

Existing legal and regulatory frameworks in the extractive sector is believed to have contributed to creating a more favorable profit-oriented business environment for investors in the extractive sector, contributing directly or in a roundabout way to overwhelmingly reducing institutional capacity in the host developing countries; driving down norms and standards in critical areas of social and economic development and the protection of the environment in many developing countries where mining and oil and gas development activities take place.

There are also some fundamental ethical and ecological issues and their consequences facing extractive sector development that must be addressed to prevent disasters that could destroy the environment and threaten all life forms, and deal with the negative consequences that have already occurred bearing in mind how they can be remedied.

Climate change has been at the forefront of international debate for quite some number of years. However, it was only in 2007, that the international community reached a scientific consensus that human induced climate change is unequivocal, and is largely a result of rising levels of Greenhouse Gas (GHG) emission mainly caused by human activity.

For instance, fuel which is the end product of oil and gas sector of the extractive industry is credited as the cause of 6 per cent of greenhouse gas emissions into the atmosphere, which is just the direct impact when the end use is transportation and heat generation and it further fast-tracks climate change and pollution by releasing by-product gases into the atmosphere during use.

Climate change though unselective of state territorial boundaries as it has already exerted severe negative impacts on people and ecological zones across the world, affecting the rich and the poor in both rural and urban areas throughout the length and breadth of the world. However, there is overwhelming evidence that the social and economic costs of climate change are disproportionately borne by those living in poor developing countries and mostly where the extractive companies are operating their business or where they have operated and subsequently abandoned.

In the poor communities in the developing countries where extractive companies operate, there is overwhelming evidence of the negative effects of climate change on poor communities and their livelihoods to be witnessed first-hand. There is increasingly negative impact of climate change on the eradication of poverty, the realization of sustainable development, and the full enjoyment of human rights by all.

Further, the world community but then again most especially the developing countries where the extractive industries operate are confronted with a growing number of disasters caused or aggravated by climate change allegedly linked to the activities of the extractive companies. The effects of climate change evidenced by researches disproportionately affect the vulnerable and poor people in developing countries through the destruction of their sources of human sustenance.

In the earlier periods for instance, rainfall was predictable just as seasons and other climatic conditions. Flooding which is associated with heavy rainfalls is also not a new phenomenon but as witnessed in the recent past, sadly enough however, that has become a thing of the past. These days, unpredictable heavy rainfalls and their associated flooding have now assumed a more severe and worrisome dimension from what used to take place in the past. These are some of the effects of *Climate Change* caused by environmental degradation from extractive industries activities and its' attributed extraordinary rise in extreme weather conditions such as higher-intensity hurricane, tsunamis and heavier rainfalls across the world. Climate change will continue to increase the frequency of heavy rainstorms, putting many vulnerable people and communities at indescribable risk of devastation from floods.

The developing countries where the extractive companies operate have become especially vulnerable to climate change because of their systemic poverty, geographical location, dependence on rain-fed agriculture, their living conditions, and the interference to their accustomed systems of survival. Though poverty already exists in these developing countries regardless of climate change, however, climate change is creating a new brutal force which deprives the people living in poverty of their ability to stabilize, improve and reverse their helpless situation.

The negative impacts of climate change affect both the poor and rich, but it is the poor people who suffer mostly the sustained and protracted deprivation of resources, capability and power, and this encumbers their choices and security.

The impacts of climate change on global meteorological systems are widely recognized to include increasing occurrence of extreme weather events, heavy and erratic rainfall, flooding and silting of rivers, sea level rise, glacial melting and retreat, sea-ice shrinking, and the contraction of snow cover, permafrost defrosting and drought that could certainly continue to disrupt many nations economy. The impacts of climate change on affected communities are multifaceted and the implications far-reaching.

For instance, according to the United Nations Human Development Report (2007), between 1990 and 1998, 94 per cent of the world's 568 major natural disasters and more than 97 per cent of all natural disaster–related deaths were in developing countries. People living in poverty are often vulnerable and marginalized within their societies due to poor-quality housing, overcrowding, and a lack of alternative livelihoods. As a result, they are more exposed to the impacts of natural disasters where many people lose their lives, most lose their dwellings and crops, and their water sources are polluted. The increased frequency and intensity of natural disasters, accelerated by climate change, mean that those living in poverty do not have the time or resources to adequately recover from one disaster before they are hit by the next.

In addition, climate change hampers access to clean, safe water in many communities in many developing countries where already inadequate water supply exists due to drought and as a result of salt water invading the soil in low-lying coastal areas, poisoning freshwater wells.

According to the Intergovernmental Panel on Climate Change (IPCC), in Africa alone the population at risk of increased water stress due to climate change is projected to be between 75 and 250 million people by 2020, rising to 350-600 million people by the 2050s if drastic action is not taken to change course.

The United Nations Human Development Report (2007) indicates that on the health front, global warming and the increase in frequency of extreme weather phenomena such as the extreme heat or cold associated with erratic temperature changes have serious implications for sanitation conditions as climate-sensitive diseases such as cholera and diarrhea, spread through water and via vectors such malaria and protein-energy malnutrition are among the largest global killers that caused more than 3.3 million deaths globally in 2002, with 29% of these deaths occurring in Africa.

When Greenhouse Gas pollution and smog that have severe impact on respiratory diseases is added, there will be higher morbidity and mortality rates.

In the poor communities in the developing countries where extractive companies operate, there is overwhelming evidence to be witnessed first-hand of the effects of climate change on poor communities and their livelihoods. There is also evidence of increasingly negative impact of climate change on the eradication of poverty, the realization of sustainable development, and the full enjoyment of human rights by all.

The poor people living in areas where extractive companies operate and who are also affected by climate change experience food and water scarcity, eroded livelihoods, most children and young women not going to school; instead they help their families in earning additional income through finding employment in the towns and cities. Such sacrifices have consequences on the ability of many of these poor people in developing countries where extractive companies operate to move out of poverty. This may likely with all probability manifests to crises at national, regional and global levels, undermining the social and political stability necessary for sustained human and economic development, and further undermine the ability for adaptation and resilience in these countries due to increased tensions over accessing diminishing land, water and food resources.

Though climate change imbues humanity with devastating prospects of disaster and ruin, at the same time this can also be seen as an unprecedented opportunity for the stakeholders in the extractive sector to cooperate as one family to remediate the problems to ensure the wellbeing and survival of both current and future generations.

No doubt, from all indications, the extractive companies must genuinely recognize the negative impacts of their activities and accept responsibility for the protection of the rights of people living in developing countries where they operate. They must accept their contribution to ecological debt to the international and local host communities, and must take-on major obligations to lead in the efforts to reduce Greenhouse Gas (GHG) emissions influence on climate change resulting directly from their activities of extracting and processing natural resources, which is already at crisis point. This will contribute to the achievement of the much bandied sustainable development.

It is imperative therefore that the extractive sector begin to think of climate change in terms of its impact on people, and its social, economic and humanitarian implications and the entire stakeholder's groups must recognize and respond to the scientific evidence and indications provided by experts to prevent natural resources which is supposed to be beneficial to people from turning against humanity that it is supposed to benefit.

In undertaking this responsibility, the extractive companies must have the strategy and courage to engage other critical stakeholders to build a strong alliance to protect the rights of host government and communities as well as the global environment and the same time providing the impetus to promote common good of their extractive activities for all.

They should contribute funds, technology and capacity enhancement that are needed by developing countries to address climate change and ensure they cultivate the strategies towards achieving sustainable development.

The strategic actions taken to address climate change must adequately take into account, along with broader environmental concerns, poverty and general vulnerability to encumber the chances of running the risk of deepening spiral discrimination.

For instance therefore, as the frequent severe rainstorm associated with climate change cannot be stopped naturally, unless through reduction in environmental degradation from extractive industry activities which contribute to climate change, preparing for the floods from the onset is more likely to be most imperative than managing the aftermath of the flooding. Conversely, other measures may come in the form of effective climate justice system that ensures the polluters pay for the cost of environmental remediation and reversal of the negative consequences of their economic activities.

The participation of local host communities, with support from other key stakeholders, in decisions regarding the management of the natural resources in their communities and engaging the extractive companies in such projects ranging from renewable energy to health, education, and controlling water in areas that have too little (or too much) among others, that could add value to their lives, will highlight the strength and resilience of local communities in taking charge of their own lives and planning for a safer future.

The consequences of greenhouse gases are well-known if not drastically reduced. Therefore, all key stakeholders must work together to make substantial reductions in carbon emissions for the sake of everyone, more especially for the sake of the poor and other vulnerable population. In this regard therefore, there must be a fair and effective climate agreement to ensure substantial and efficiently delivered funding to help communities address potential hazards of the extractive sector activities and prepare for future climate changes.

This type of intervention will provide those people in developing countries where the extractive companies operate, especially those who are traditionally most marginalized such as women, indigenous communities, disabled people, and the civil society organizations that advocate for the protection of the interests of these people and the environment, and the governments that have the responsibility to protect life and property the necessary capacity to play significant roles in the abatement of the negative impacts of climate change.

Another area that must be focused on as cross-cutting issue is gender inequality in natural resources management. The term gender refers to socially constructed roles and social expectations of individuals, based on their sex as male or female.

The existing gender relations favor men, and the stereotypes based on gender that are deeply ingrained in society see women's status as traditionally low, and women are mostly excluded from political and cultural life even when they can perform in the same way if not better than a lot of men.

Gender inequality brings about such disparities as unequal opportunities in education, unequal opportunities in employment, unequal rewards for similar work, unequal access to various kinds of productive resources, and an asymmetrical opportunity for women to participate in and influence decisions that shape their development process.

Women in all sincerity are technically disempowered. This disempowerment of women and girls through gender inequality has contributed inordinately to the rampant discrimination and egregious sufferings of the female gender from the direct or circuitous actions of host community's leaders, extractive companies and the host government alike.

Thus many women work in low-paying jobs and are burdened with unpaid domestic work and children upbringing. Most women are to a lot of degree reliant on their husbands' incomes and when their husbands are negatively affected by the actions of natural resources development companies, it affects the women exponentially because women's ability to provide nutrition for themselves and their families for example intersects with their capacity to control resources and make decisions in the household. Their status as women in a world of absolute gender based discrimination against women as a result of societal dictates contributes to their inability to provide healthy food for themselves and their families.

The roles and activities of the Bretton Woods institutions, the World Bank Group (WBG) comprising the World Bank itself, International Finance Corporation (IFC) that provide funding for the investors; the Multilateral Investment Guarantee Agency (MIGA) that provides insurance coverage for these investments, and the International Monetary Fund (IMF) as well as the home countries of the extractive companies in mismanagement of the extractive sector in developing countries where solid minerals and hydrocarbons are developed in exceptionally unsustainable manner are also reviewed in this book.

In meeting the development challenges and reducing the rancorous conflicts in the mineral-rich developing countries, this book also discusses issues regarding the necessity for extractive companies to ensure the legality of their activities, and thus raise the question of their corporate social responsibilities and their legitimacy.

The book also delves into some possible unequivocally legitimate activities that will lead to sustainable natural resources development, bearing in mind the need for multinational corporations to rethink their involvement as their activities have allegedly negatively impacted humanity, especially the most vulnerable groups.

In fact, this book is making a clarion call for paradigm shift towards upholding the doctrine of sustainable socio-economic development in the extractive sector. This clarion call is germane for the fact that traditionally, there has been a high level of distrust in the natural resources sector that some extractive companies with good plans for the development of their host and impacted communities find out that the level of trust by the communities fall far below their expectation, thus leaving them to struggle to establish legitimacy and understanding.

This book ultimately aims to improve the widespread unsustainable approach to the management of the natural resource endowment of developing countries. This is to make sure that the activities of all key stakeholders in the extractive sector and revenues generated from natural resource extraction contribute optimally to achieving sustainable economic growth and livelihoods enhancement, thus reversing the age-long agonies of the host communities and states. This will be achieved by identifying and analyzing the historical challenges overwhelmingly present in the extractive sector that influences the activities and behaviors of key stakeholders in the extractive industries. In addition it will provide a more plausible platform to tinker about the framework of engagement of all key stakeholders on clearly identified entry points.

CHAPTER TWO

Major Stakeholders in the Extractive Sector

As in every economic sector there are major stakeholders visibly present that influence or are influenced by the activities of that economic sector, whether positively or otherwise and the extractive sector should not be any different.

In the case of the extractive sector, the major stakeholders identified in this book include the government, the development partners, the extractive companies, the host communities, the civil society, the media, and the employees among others. This section of the book reviews the responsibilities and activities of these stakeholders and how they affect the extractive industry and how the extractive industry affects them too, starting with the government.

THE GOVERNMENT

The government is a major stakeholder in the extractive sector. In majority of developing countries, the government owns the natural resources in their various countries. They participate actively in the development of these natural resources through their ministries of petroleum resources and the solid minerals.

The government enacts the laws governing the extractive sector, enters into contracts with extractive companies; regulates the industry and issues licenses to the extractive companies to operate and also monitors the operation of the sector.

The government generates revenues from the extractive sector through taxes, bonuses, royalties and through participation vide joint venture agreements, production sharing contracts, among other evolving arrangements. The government is responsible to manage the revenues derived from the extractive sector in the interest of the general public and also provide the needed oversight functions. To effectively carry out these responsibilities, the government requires some resources of vital importance which include human and finance.

The government needs to have in place available requisite technical skills necessary to enact appropriate legislation and regulate the extractive sector. The government needs to have in place, qualified staff that understand the workings of the extractive industry, have the capacity to apply the laws and regulations guiding the sector, have the ability to monitor industrial activities and carry out audits work.

The government needs qualified staff that can develop appropriate fiscal regime for the extractive sector, calculate appropriate signature bonuses, royalties, taxes, payments in kind and transit revenues, contributions, earnings, holdings and withdrawals/distributions, including the budget; investment rules, regular independent financial audits.

The lack of qualified staff with industry-relevant skills will portend government's inability to take full advantage of the extractive sector. This is likely to manifest to communal conflict, conflict among the stakeholders and revenue leakage and mismanagement through corrupt practices.

The government has in most extractive jurisdictions created problems in the host communities, by not carrying them along throughout the process of engaging the extractive companies even up till revenue management. This attitude results in poor harmonization of institutional capacity, community engagement and decision-making within a state system that contributes to encumbrance in extractive sector governance. Focussed capacity building efforts do not necessarily solve this problem, as governments do break rules of good financial practice, even their own laws, in order to move money about to in order to service *ad hoc* needs.

Poor harmonization of institutional capacity contributes also to continued fragmented individual and institutional roles and responsibilities, which hampers the ability of government to coordinate a sustained effort to improve extractives governance as well as depriving the country of maximum economic and social benefits from the extractive sector.

Though the legislative arm of the government has constitutional oversight responsibility for holding the executive in check, most of the legislators currently have little capacity to fully understand and analyze the activities of the extractive sector therefore, this negatively affect their ability to influence policy in the extractive sector.

The lack of government's ability to provide a consistent quality of basic services such as employment opportunities, healthcare, education, road maintenance or water supply and sanitation improvements, leads to additional (and unwelcome) pressure on extractive companies to provide such benefits and services in their areas of operation through corporate social responsibility programs.

The onus therefore lies on the government to ensure that there is progressively sustained development that will translate natural resources generated wealth into wealth for the greater population to reduce impacts of poverty on the communities, reduce unmet commitments on royalty and tax obligations, environmental destruction and, even worse, significant social discontent and potential for further conflict.

DEVELOPMENT PARTNERS

Development partners (DPs) are critical stakeholders in the extractive sector as they strongly influence governance mechanisms in the extractive sector through the encouragement of harmonization investment regimes, supporting the processes involving the drafting of engagement rules, procedures and regulations for effective collection and transparent management of the sector and the revenues generated from the sector.

The DPs are known to influence the host states to enact extractive governing laws, adopt strategies and policies favorable to them and the investors, instead of the host states. This is especially true when the bilateral DPs share the same country of origin with the extractive companies, or is diplomatically influenced by the home state of the extractive company and this in no small measure creates potential for conflicts of interest.

Many extractive jurisdictions have suffered reversed economic growth and social conflicts, including increased unemployment, job losses, increased poverty, and communal crises, from the Structural Adjustment Programs (SAP) recommended by the DPs. Further, SAP encourages less social investments in environment, removal of subsidies in the extractive sector, subtler but yet very latent discouragement of budgetary deficits, obviating necessary infrastructural development and social security provisions for the less privileged. Another issue is the rivalry among the DPs for engagement with the host government decision-makers which tend to weaken their collective resolve to influence provision of policy advice and technical support, especially when the commitment is focused at the highest level of government.

EXTRACTIVE COMPANIES

Extractive companies have been extracting natural resources from beneath the earth in the jurisdiction of developing countries for so many years now, some dating back to colonial days before the country became independent from the colonial government. Most of the concessionary agreements from the colonial governments were such that favored the host communities far less than the extractive companies, who went about plundering the natural resources without any challenges from the host communities. In this regard, the extractive companies have exerted a strong influence over the governance of the extractive sector and while this is understandable to a certain extent, their aim was to maximize profits and protect their investments without any thoughts or understanding on their own part on how to protect the downstream implications of their operations, which include tales of woeful environmental despoliation and fatigue, bionetwork destabilization, and human rights abuses that consist of unlawful arrests and illegal detentions, forced-labor, forced-displacements or arbitrary relocation of residents, rapes and extra-judicial killings, meted out to local residents of these natural resources development jurisdictions.

Some of the extractive companies have been allegedly linked to the utilization of some of the income they realized from natural resources development to fuel and sustain conflicts; engage in illegal arms trafficking through clandestine operatives who supply arms to militaristic parties; acquisition and usage of weapons of mass destruction by warring groups and money laundering that manifests to mordant poverty in the communities and causing untimely death of innocent citizens.

There is the story of a legal action by Ok Tedi and Fly River people in the South West Province of Papua New Guinea against B.H.P. Billiton of Australia, the Chanaral and Huasco cases in Chile, the case against Freeport McMoRan Inc., by the people of Irian Jaya, Indonesia; the Omai Gold mine case in Guyana against Cambior Inc.; the wars in Liberia, Sierra Leonean "Blood Diamonds" or "Conflict Diamonds".

But now as these former colonies attained their independence and establish their sovereignty over their natural resources, there began to emerge numerous stakeholders with varied interests in the extractive sector. With these interests the extractive companies have faced different challenges in their areas of operations. Some of these challenges have included expropriation of the companies, leading to loss of assets by the investors; well celebrated cases such as the ARAMCO case.

There have been and still are cases in the Niger Delta of Nigeria against Royal Dutch/Shell, Chevron, AGIP, ExxonMobil, ELF, and Texaco. There have also been cases against Texaco Petroleum Company in the Amazon rainforest basin and Total S.A. and the Myanmar Oil and Gas Enterprises of Burma (Myanmar) in Myanmar among others.

With all these cases, the extractive companies should now realize the need to have in place a clear incentive to build and foster relationships with a wider range of stakeholders (notably, communities, trade unions and development partners). This will ensure the existence of adequate security for the protection of their operations. They will also need to exert extra influence over government to observe consistency in respect of royalties and taxes.

EMPLOYEES' UNIONS

Another critical stakeholder group in the extractive sector is the employees or the organized labor through their unions. This group of stakeholders do influence the success or failure of the extractive sector. Several trade union organizations for one reason or another have downed tools in protest of government policies (not only in the extractive sector), or industry or company's response or lack thereof to specific union demands.

The world is aware of the downing of tools by mine workers in South Africa after employers refused the upward review of their salaries. In Nigeria, the Petroleum and Natural Gas Senior Association of Nigeria (PENGASAN) and National Union of Petroleum and Gas (NUPENG) have successfully disrupted the supply of petroleum products to end users, causing fuel scarcity in the country for weeks that eventually negatively affected other economic sectors.

The employees' unions need to be taken serious as they have significant opportunity to influence the level of industry production. The problem of addressing the rights of workers, so often ignored by the employers in the extractive sector, in the promotion of employment and livelihood opportunities, needs to be tackled through appropriate representation of trade unions which in best practice is the true vehicle for reduced industrial actions in the extractive sector.

THE CIVIL SOCIETY AND NGOS

Civil Society Organizations (CSOs) and Non-Governmental Organizations (NGOs) are critical stakeholders in the extractive sector who have campaigned effectively to point out the need to make the extractive sector more transparent and on the need to change the law and regulations to reduce the plight of some mining-affected communities. They have engaged the government and the extractive companies using industry experts from within the country and internationally to produce high quality analytical reports on the extractive sector. CSOs and NGOs have also engaged in numerous debates over extractives sector governance with politicians at public meetings and on national media outfits. However, a great number of them lack relevant capacity to understand the workings of the extractive industries hence some of them are considered to be distracters by both government and extractive company's practitioners.

Some of the CSOs and NGOs are members of networks engaged in strengthening the capacity of the communities and the general public to effectively engage both the government and the extractive companies and to participate in decisions regarding resource governance.

On some occasions, the CSOs and NGOs have encountered hostility from local populations who are desperate for development investment and fearful that adverse publicity will drive foreign companies away. Also, some CSOs and NGOs campaigners have been illegally arrested by government forces for daring to speak up against corruption, environmental and human rights abuse prevalent in the extractive sector.

Another problem of the CSOs and NGOs working in the extractive sector is that, like the media, they are heavily dependent on contacts in government to supply them with information e.g. copies of mining license agreements, tax regimes and budget documents with which to carry out their advocacy work. This dependency again serves as a disincentive to critical analysis that might alienate these sources.

There is need to strengthen their capacity through developing a "Think Tank" where information are gathered and professionally analyzed that will inform the platform for the CSOs engagement in the extractive sector, so their contribution should not be dismissed as irrelevant and frivolous.

All in all, the civil society organizations have significant part to play in ensuring the reversal of negative impacts of extractive sector development and lifting the natural resources curse from the public by organizing themselves into advocacy and campaigns coalitions involving such organizations as the labor unions, professional bodies, religious groups, the media organizations, and educational institutions, among others and using such platforms as the electronic, print and social media, town hall meetings and rallies.

THE MEDIA

Like in every other economic sector, the media, comprising the print and electronic, is another critical stakeholder in the extractive sector. The media is the public watchdog responsible to bring information about the operations and activities of the extractive companies to the living room of the citizens. Having access to such information has been a huge challenge because most agreements between the host governments and the extractive sector are treated as national secret documents that cannot be released to the public, for fear it will be compromised or it will trigger public discontent.

Apart from accessibility to concession documents the media in the extractive sector is encumbered by limited knowledge of industry operations, making their reportage in the sector very incomprehensive. Further, in most of the mining jurisdictions, the media practitioners lack access to operations site and can only get there when conveyed by the extractive company with its helicopter or boats. That being the case, before some media outfits can get to areas of oil spill in the high sea or in the mountains where the resources are extracted, a lot of cover up would have taken place and in fact, the media personnel could be prohibited or restricted by the extractive company and government on what must be reported to the public.

Articles on mining governance appearing in the press vary considerably in quality, and critical analyzes are often dismissed by government spokespersons as the work of opposition activists, pursuing political as opposed to a developmental agendas. Some issues no matter how justifiable they are and especially those bearing on the integrity of senior government officials are regarded off-limits even to what may be described as impartial reporting.

Government tends to badge all criticism as 'unpatriotic' and take the position that you're either for us or against us. To improve their quality of work the media which is the fourth estate of the realm need to be empowered to ensure timely access to information and their capacity built to ensure availability of information and improved quality of analysis and reporting of the available information.

COMMUNITIES

The strong sense of community ownership over land and its resources leads to an equally strong sense of entitlement to benefits from outside agencies using 'community' land for their own profit. This sense of entitlement is a product of decades of poverty and isolation, but makes it difficult for companies to build constructive relationships with the rural population.

The host communities in developing countries of the world living in and around where natural resources are extracted are the major victims of the activities of the extractive companies. There are confirmed tales of woeful environmental despoliation and fatigue, bionetwork destabilization, and human rights abuses consisting of unlawful arrests and illegal incarcerations, forced-labor, forced-displacements or arbitrary relocation of residents, rapes and extra-judicial killings, causing the untimely death of innocent citizens through environmental pollution to mention a few.

Being the stakeholder group most affected by extractive activities, local communities' capacity to influence policy and governance of the extractive sector ought to be high but unfortunately it is currently on a minimal spectrum.

The host communities and extractive companies have fundamentally frosty relationship due in part to:

a. The disruption caused by the extractive companies operations on the land which is the source of livelihoods of the host communities;
b. Low employment of the host community population due to lack of industry required skills;
c. Communities' recurrent fear that their political representatives are more interested in taking commissions from the extractive companies than in championing the course to protect their interests;
d. The undeniable fact that historically, communities have seen few or non-existent sustainable direct benefits from extractive sector.

The government and the extractive companies have caused numerous intra and inter-community tensions (e.g. grievances against chiefs, inter communal clashes and uneasiness due to the influx of non-community migrants in search of jobs in the extractive companies).

In many developing countries where natural resources are exploited, education and cultural norms do not encourage women to speak in public like the men, yet women play a key role in the sustenance and development of humanity and the human society. Even though they constitute around 51-53 per cent of the population and account for about 90 per cent of agricultural labor, doing most of the transporting and processing of farm and forestry products but as farmers, they can observe the changes that are taking place in their environment, such as the impoverishment of the soil due to climate change in the fields where they work, but because they are excluded from public debate, it is hard for them to help find solutions to their very own problems. All these among other things exacerbate women's vulnerability and poverty. Prohibited from speaking out in their communities, women are thus denied the right to sustainable livelihoods.

Women play very critical role in human development however, their participation is encumbered by such factors as: lack of capital; socio-cultural and institutional discriminatory practices and limited educational opportunities.

Closing gender gaps in the extractive industry through strategic intervention beyond the current palliative and soporific approach is generally attractive, not only because it will improve the lives of women and tends to raise their relative status, but also because it will generally propel the advancement of sustainable socio-economic development to the benefit of the general public not just the female gender. Therefore, taking significant action to mitigate the unequal gender relationship will have positive outcome for the women in particular and society in general.

Though enabling laws and treaty commitments should have resolved this issue but lack of seriousness in implementing these laws and treaties forecloses their effectiveness. It is in this regard that well-thought through strategies should be institutionalized to ensure reduction in the rate of suffering of the female gender.

To make progress along this line, the above issues encumbering gender equality should form part of the guiding light when dealing with gender inequality as cross-cutting issues in natural resources management.

CHAPTER THREE

Some Impacts of Extractive Sector Development in Selected Developing Countries

The agony of host and impacted communities especially and countries at large in where natural resources development are undertaken in African, Asia, Latin America, and the Oceania is overwhelming as this development has always for the most part resulted to negative consequences for the hosts. Both the development companies and the host governments are unfortunately mostly fingered as the perpetrators of this oddity.

The challenges of the management of extractive sector in developing countries are the same all over the globe. It is known facts which have been well documented across the globe that natural resource-rich developing countries have the worst development performance indices than those with smaller endowments or developed countries with huge deposits of natural resources hosting extractive companies and this phenomenon is what has become known in development phraseology as "*resource curse*".

These developing countries where the extractive sector provide the main source of revenue for the government in all fairness when compared to countries that earn their revenues from agricultural commodities and manufacturing industries seem to be the most economically troubled, parading the most authoritarian leadership, seem to be the most conflict-ridden in the world, suffer from abnormally high rate of poverty, poor health care regime, widespread diseases, high rates of child and maternal mortality, low life expectancy, and poor educational performance. This is summarily attributed to the linkage effect of fragile public institutions, authoritarianism, corruption, conflict and primitive accumulation of wealth through collection of bribes and contract inflation that are prevalent in these developing countries.

Sustainable extractive sector management can be achieved through a set of strategic approaches designed to improve the level of transparency and accountability of governments and extractive companies during the contracting, licensing, exploration, extraction, revenue generation, allocation and application of the financial resources from the sector for socio-economic growth and development. The mechanisms to achieve it include policy instruments, financing mechanisms, rules, procedures and norms that regulate the activities and processes of the extractive sector. The three key elements to achieve this are adequate environmental laws and regulation with proper enforcement mechanism; good corporate social responsibilities practice by companies; transparency and accountability with participation of communities and civil society organizations which are effectively lacking in most of the developing countries where the extractive industries operate.

This book now looks at some selected extractive jurisdictions with a view to highlighting some seeming negative consequences of unsustainable natural resources management therein, beginning with Angola.

ANGOLA

In Angola, where diamonds were first discovered in the north-eastern Luanda Province in 1912, and where in 1917 a mining company, the *Companhia de Diamantes de Angola* (DIAMANG) was formed to begin mining operation mainly in Luanda Province and in the Cuango Valley to the west, diamond mining has provided substantial amount of financial resources for the colony and was equally the source of nemesis to majority of the citizens and mine workers and their families.

For instance, it was alleged that, in 1947, about two-third of DIAMANG 17,500 African workers were believed to have been acquired through forced labor and their payment whether in cash or in kind typically averaged about 830 angolars (US$25) a year[5].

Then, when the war of independence started in Angola in 1961 with three different armed guerrilla movements namely *Movimento Popular de Libertação de Angola (*MPLA), the *Frente Nacional de Libertação de Angola* (FNLA), and the *União para la Indepêndencia Total de Angola (*UNITA) led by a 31 year-old Jonas Savimbi, diamond became a source of misery for the citizens of Angola.

The Soviet bloc, the West, Cuba and South Africa supported Angolan war of independence. However a 1991 cease-fire eventually led to elections the following year. But when Jonas Savimbi who was favored to win the election lost in the poll, the war resumed. By now, the support of their Cold War allies was not enough for the warring MPLA and UNITA therefore they focused greater attention on oil and diamonds from Angola as domestic sources of support for their war efforts. UNITA had begun exploiting diamonds in the 1970s, and in 1984 it

---

[5] Basil Davidson, *In the Eye of the Storm: Angola's People* (Garden City, N.Y.: Anchor Books, 1973), 70.

overran key diamond areas in the Cuango Valley, ostensibly exporting about US$4 million worth of diamond gems that year. By 1993, UNITA had taken over and was in full control of most of the best diamond areas. By 1996, it was believed that UNITA was exporting a staggering US$1 million worth of diamond gems every day. The government itself on the other hand, depended on the lucrative sales of oil[6].

At the end of the 37 years of warring, it was believed that as many as 300,000 Angolans had died in battle, and hundreds of thousands more had died indirectly. Millions were displaced and the country's infrastructure, seriously underdeveloped at independence in 1975, had almost been completely destroyed. It was estimated that 200,000 people had been disabled by land mines; more than two thirds of the population lived on less than a dollar a day, and three children out of ten died before their fifth birthday[7]. Amidst the carnage, conflict continued, and the commercial channels funnelling UNITA diamonds out to world markets and bringing weapons back into Angola were among the few things that were actually left running in Angola.

UNITA was believed to have always controlled 60–70 per cent of Angola's diamond production, generating US$3.7 billion in revenue, which facilitated their war effort. UNITA's diamonds,' as was reported in the United Nations, 'reached the international markets through a worldwide diamond industry that operates with little transparency or scrutiny from the international community.[8]

The Angolan government expelled an estimated 260,000 foreign citizens from Angola over an 18 month period starting from 2003 amidst brutal acts of robbery, rape, death and the administration of emetics and laxatives that induce diarrhea aimed at finding every last diamond before the aliens were driven across the border. Many of the expelled people were believed to have come from the Democratic Republic of Congo (DRC). This act of brutality and inhuman treatment perpetrated because of bloody diamond development was repeated every year against the victims.

LIBERIA

The April 12, 1980, military coup in Liberia set the stage for what will eventually give rise to the war that will be sustained by natural resources. On the night of April 12, 1980, seventeen soldiers launched a coup in Liberia that sacked the civilian government of President Tolbert resulting in the death of the president after he was

[6] Ian Smillie, *Blood on the Stone: Greed, Corruption and War in The Global Diamond Trade* (London: Anthem Press 2010) page 46.
[7] Ibid
[8] In 1999, Human Rights Watch estimated the total at US$1.72 billion.

disemboweled in the Executive Mansion. Ten days later, 13 senior members of his government were executed in front of celebratory crowds on a Monrovia beach. This coup brought Master Sergeant Samuel Doe into government as Head of State and co-chairman of the "People's Redemption Council".

Sergeant Doe as he was popularly called made himself the hub of Liberia's economic maneuvering and corruption and he personally managed the Forestry Department Authority which collected logging fees; engaged in business with Thai Army Generals who were involved in Cambodian timber deals and companies associated with private armies in Lebanon. By 1988, Doe had used up his international goodwill and subsequently, he made himself public enemy number one through his brutality, corruption and mismanagement of public resources.

Charles Taylor who had been given appointment by Samuel Doe after the coup to be in charge of the General Services Agency (GSA) which was the government agency responsible for the allotment of government property, however, was later accused of mismanagement of public funds but he escaped Liberia before a warrant was issued for his arrest.

With consistent hounding by Doe and subsequent arrest of Charles Taylor in the United States, Mr. Taylor became embittered with Doe and with the support of the Libyan President, Mummer Gaddafi, Taylor took advantage of Doe's dwindling political support from former allies and on December 24, 1989 he launched a campaign to dethrone Doe's administration. Later, Doe was captured by one of Taylor's rivals, Prince Johnson, and tortured before being murdered. However, before the end of this war about 60,000 to 80,000 people allegedly lost their life.

The bloody war did not stop with the death of Samuel Doe. To support his war machinery, Taylor embarked on exploitation of natural resources beginning first with timber, sold through consortia of foreign companies and syndicates of Liberians and Lebanese traders based in Liberia. He also turned to rubber, iron ore, and gradually more to Sierra Leone diamonds.

With all the financial resources from Sierra Leone diamonds, he was able to sponsor Forday Sanko to cause mayhem in Sierra Leone and later in Guinea where in September 2000, Sierra Leone's Revolutionary United Front (RUF) attacked several southern Guinean border towns. This cross-border attack affected Sierra Leonean refugees who had fled their country to Guinea.

Due to financial resources from natural resources many Liberians, Ivoirians, Sierra Leoneans and Guineans lost their lives; some were tortured, raped and executed in cold blood, and thousands languished in refugee's camp being starved to death.

SIERRA LEONE

In 2004, the Sierra Leone Truth and Reconciliation Commission (SLTRC) reported that the central cause of the civil war in that country had been the endemic greed, corruption and nepotism of political elites, who plundered the nation's assets, including its mineral riches… robbed the nation of its dignity and reduced most people to a state of poverty[9]. To avoid a future repeat of the carnage witnessed in Sierra Leone during the dark days of the civil war, there was a need at this stage for a detailed re-examination and refocus of the institutional frameworks of the extractives sector to strengthen Sierra Leone's government, citizens and support international development partners in ensuring that revenues are successfully managed and channeled into development and poverty reduction agendas.

The citizens of Sierra Leone for many years were denied access to the common wealth revenue generated from their diamonds exploitation. Presidents of the country like Siaka Stevens and Joseph Momoh who took over from him were described as exceedingly corrupt and tyrannical as what was foremost in their thought was how to acquire more personal wealth from the country's diamonds exploitation without recourse to the welfare of the majority of the citizens.

In April 1992, Momoh was overthrown in a military coup led by 27-year old army captain Valentine Strasser, bringing the National Provisional Ruling Council (NPRC) to power on a promise to end corruption. Unfortunately however, Strasser was engaged in a further looting of the country's diamond resources to the disillusionment of many citizens and international observers.

Foday Sankoh began his war on Sierra Leone in March 1991 after spending some time in Benghazi, Libya, learning the arts of revolutionary warfare from Mummar Gaddafi, *Revolutionary Training Camp.*

Foday Sankoh who was described as a "charismatic sociopath" created the Rebel Group the "Revolutionary United Front (RUF)", in Sierra Leone according to him with the mission to liberate Sierra Leoneans from the clutches of tyranny. However, he converted into a despotic ruler who presided over the murder and mutilation of the very civilians he purportedly assumed he wanted to liberate.

The Revolutionary United Front (RUF) rebellion began in 1991 and was supported with revenues derived from Sierra Leone diamonds which were discovered in 1930. The revenues from diamond extraction became the source of finance to a decade-long war characterized by banditry and horrific brutality that was inflicted primarily on the

---

[9] SLTRC, *Witness to truth: report of the Sierra Leone Truth and Reconciliation Commission*, Volume 2 Chapter 2, 'Findings', 2004.

civilian population. The march on Sierra Leone by the RUF rebels left thousands of women, men and children butchered and some were left without hands and feet, mutilated and disfigured physically and psychologically for life. It is believed that about 75,000 people – most of them civilians – lost their lives in the war, but the number could almost certainly be much higher.

The RUF forced young male children to be recruited into their rebel army as combat soldiers while their female counterpart was forced to become their sexual slaves. Almost half of Sierra Leone population were displaced or were refugees. The schools, hospitals, government services and commerce grounded to a halt bringing the country to the verge of total collapse as nothing was functional except in the largest urban centers. The country's diamond resources which could have been used for sustained socio-economic development to improve the lot of the ordinary citizens were used instead to finance atrocious war that robbed the potential beneficiaries and an entire generation of children of much needed development, resulting in Sierra Leone ranking dead last on one of the United Nations Human Development Index.

With the backing of the destructive Liberian warlord, Charles Taylor, who by giving Foday Sanko a Liberian base, weapons, and an outlet for whatever he could steal in Sierra Leone, facilitated Forday Sanko unquestionable military facility to unleash the act of terror on the Sierra Leonean civilian population. The RUF operational tactics was a terror-based technique characterized by horrific rapes, chopping off civilians (often small children) hands and feet.

However, all the atrocious acts of Foday Sanko and his RUF against the Sierra Leoneans and Guineans paid him off handsomely as he became the county's Vice President; in addition, the RUF was given four ministerial posts in Tijan Kabbah's government and a blanket amnesty for all RUF fighters. Foday Sanko was also made head of a new commission to oversee the country's diamond resources under a cease-fire initiated and negotiated through direct personal pressure from Jesse Jackson and American State Department officials in a United State sponsored negotiation.

This was a tragic end for the Nigerian soldiers who lost their lives in peacekeeping mission in Sierra Leone. It was a huge tragedy for thousands of family members of these soldiers who lost their loved ones. It was a tragic end for thousands of Sierra Leonean citizens who lost their lives. It was a huge tragedy for those who had their hands and feet chopped off by the RUF. What a tragic end for thousands of Sierra Leonean citizens who were displaced and or those who ended up in refugees' camp! What a tragic end for thousands of Sierra Leonean citizens who were raped by the RUF soldiers, what a tragic end for thousands of Guineans who were attack and had their homes destroyed by invading RUF rebels?

In fact the end to Sierra Leone's civil war demonstrates that as long as it is Africa, the developed countries can do anything and get away with it. It further lends credence that the preached sustainable development of natural resources for the benefit of the present and future generation is a mirage.

The RUF had demonstrated that butchery paid off. Instead of being punished for the war crimes and crime against humanity, they were rewarded, and they were assisted in the process by the most powerful government on earth-the Almighty United States of America.

At precisely the same moment that North Atlantic Treaty Organization (NATO) was spending billions of dollars to save the citizens of Kosovo in Europe from human rights abuse, much worse atrocities were being generously rewarded in the African country of Sierra Leone. Instead of going to prison for his atrocious crime against humanity, Foday Sankoh was made vice president of Sierra Leone.

As a prize for eight years of diamond theft, he was put in charge of the country's entire mineral wealth. Assistant US Secretary of State, Susan Rice, bragged at that time about how 'the US role in Sierra Leone … has been instrumental. With hands-on efforts by the president's special envoy, Jesse Jackson, Ambassador Joe Melrose, and many others, the United States brokered the cease-fire and helped steer Sierra Leone's rebels, the Kabbah government, and regional leaders to the negotiating table[10].'

THE CONGO

Diamonds almost from the day they were discovered in 1907 has played a central role in the Democratic Republic of Congo (DRC) what could be professionally described as unhappy encounter with modern history.

The DRC, once known as the 'Congo Free State' during which period this central African Nation was regarded as the personal fiefdom of Belgium's King Leopold II has continued to be polluted by blood diamonds dating back to its colonial rule by King Leopold II. In his conceited ambition to find a colony in Africa, the greedy and overambitious King Leopold II, in 1876, hosted a grand 'Geographical Conference' in Brussels, bringing together the world's foremost explorers, geographers and humanitarians where he announced to his audience that the purpose of the gathering was to deliberate on how "To open to civilization the only part of our globe which it has not yet penetrated. To pierce the darkness which hangs over entire peoples is, I dare say, a crusade worthy of this century of progress".

---

[10] Ryan Lizza, 'Where Angels Fear to Tread', *the New Republic*, 24 July 2000.

To achieve his mission, he first of all set himself up as a patron of exploration and then as a great humanitarian, determined to rid the African continent of Arab slavery, and to civilize the heathen. He then created a philanthropic organization "the International African Association", which in later years turn out to be the front for odious carnage against the people of the central African nation.

King Leopold II engaged the service of the great African explorer, who had found the long-lost missionary, David Livingstone in 1871, Henry Morton Stanley, to plant his flag which was a gold star on a blue background, wherever he could. In fact, the discovery of diamonds became the underpinning on which a cursed instead of blessed DRC emerged.

Ivory, tropical hardwood, palm oil and minerals were extracted by companies that were given generous concessions over vast territories. A private army, the *Force Publique*, was created to enforce Leopold's II order and to ensure adequate supplies of labor for the building of roads and a railway into the interior. Villages were given quotas for the production of rubber from wild vines in the country's forests. Failure to meet targets brought down the wrath of the *Force Publique*, which raided and destroyed entire villages, killing men, women and children. Many people lost their lives and others had their hands and feet chopped off. A Belgian government enquiry in 1919 estimated that from the days when Morton Stanley began to plant Leopold's flag, the population of the Congo had been reduced by half. According to the author of the book "*King Leopold's Ghost*", Adam Hochschild, the first people died from outright murder – war made by the *Force Publique* and others, on innocent civilians. More died from starvation, exhaustion, exposure and the spread of disease. The population who died or fled during those first years of Belgian influence is estimated in several studies at ten million people, and in one at a staggering thirteen million[11].

Copper, zinc, gold, tin and diamond mining in the Katanga region was very important to the allied forces during World War I and even more important during World War II when most of the uranium used to develop the atomic bomb that destroyed Hiroshima and Nagasaki came from mines in the Congo. Unfortunately, all colonial production was believed to have been accomplished with forced labor.

Even, it is believed in some quarters that the first political disturbance that occurred seven days after DRC attained independence in 1960 erupted around minerals. The Army mutiny just within seven days of attaining independence led to the secession of mineral-rich Katanga province. The Prime Minister, Patrice Lumumba, and the president were sacked by the newly-appointed army chief, Joseph Mobutu, and this was followed by

[11] Adam Hochschild, *King Leopold's Ghost* (Boston: Houghton Mifflin, 1998), 44.

the assassination of the ousted Prime Minister[12]. Five years of rebellion, secession, reintegration and confusion followed, until October 1965, when Joseph Mobutu stepped in again, this time taking control decisively and for good until 1997 when his very corrupt regime collapsed under the weight of its own corruption.

When a group of Congolese rebel organizations announced the formation of the Alliance of Democratic Forces for the Liberation of Congo-Zaire (AFDL) at the end of 1996, Rwanda and Uganda assisted with weapons, troops, logistics and cash.

With support from Zambia, Zimbabwe, Eritrea and Tanzania who provided weapons or staging bases for Mobutu's antagonists, Mobutu's army, which was so corrupt and rotten to the core, fell back without engaging the rebels, looting as it retreated. The AFDL, led by long-time rebel leader Laurent-Désiré Kabila, took the diamond towns first: Kisangani in March 1997, Mbuji-Mayi in April, and then Lubumbashi, in May, Western embassies were urging quick takeover and contacted Kabila to crave for a more rapid advance into Kinshasa, fearing a complete power vacuum and a breakdown in whatever law and order was left if Kabila did not move fast. Amidst the chaos, the Father of the Nation flew out. His exit was arranged by his last diamond co-conspirator, Jonas Savimbi, who sent an ancient Ilyushin cargo plane to airlift Mobutu and his grasping family to their final exile.

Laurent Kabila, whose troops marched into Kinshasa on 15 May 1997, had begun his fighting against Mobutu's army in the early 1960s. In order to finance his march to Kinshasa, Kabila sold diamond and other mineral concessions left, right and center, signing contracts with companies as eager for profits as himself. One such arrangement gave America Mineral Fields important cobalt and diamond concessions. A lot of cash and Kabila's use of a company plane helped clinch the contract.

After taking over power, Kabila abrogated that deal and many others, including all the mining agreements made by Mobutu in his final scramble for cash. Kabila in 1998, embarked on a numerous 'reforms' in the diamond sector, with the sole purpose of increasing his control and the revenues that would be realized from the sector. He banned foreigners from diamond mining areas and he required traders to pay a US$25,000 performance bond along with their taxes in advance. He required dealers to sell through a Kinshasa bourse, which had a membership fee of US$3 million. He halted the sale of diamonds in anything but the local currency, which was all but worthless. The outcome was the drop in the sales of diamonds through official channels by one third from US$451 million in 1998, to only US$290 million in 1999. This was followed by other exports.

---

[12] The true story about the death of Prime Minister Patrice Lumumba is one of the most kept secrets in the world till date. Some people point accusing finger to the American Central Intelligence Agency, while some others finger the Belgian government.

Kabila was shot dead, under circumstances that to this day remain unclear and the son of Kabila, Joseph, became President of the Congo on the death of his father in January 2001.

DRC's history is both tragic and bloody due to exploitation of natural resources. A 1919 report estimated that some ten million people had died during the first 40 years of Belgian association with the Congo. In March 2001, an American humanitarian organization, the International Rescue Committee (IRC), released a report that was eerily reminiscent of a 1919 enquiry conducted by the Belgian government and the IRC report estimated that in a 32 month period between August 1998 and March 2001, 2.5 million more people died than would have been the case under normal circumstances.

Most of the deaths resulted from disease and malnutrition brought on by war financed by revenues from natural resources exploitation, 350,000 people were killed in conflict. In April 2003 they issued a new report, boosting the number to 3.3 million, and in January 2008 they revised the numbers upward again, to 5.4 million deaths – the worst human calamity since World War II[13].

IRIAN JAYA

Irian Jaya is an equatorial island in the province of Indonesia. This island is located on the western part of the island of New Guinea. It covers the Bird's Head (or Doberai) Peninsula and surrounding islands. Irian Jaya is about seventy-eight miles and is an exceptional sanctuary imbued with biodiversity comprising tropical rainforest with the tallest tropical trees, marsupials (including possums, wallabies, tree-kangaroos, cuscuses), other mammals (including the endangered Long-beaked Echidna), many bird species (including birds of paradise, cassowaries, parrots, cockatoos), the world's longest lizards (Papua monitor) and the world's largest butterflies. The island is also estimated to have over 16,000 species of plant, with broad waterways and wetlands that make a perfect habitat to salt and freshwater crocodile, tree monitors, flying foxes, osprey, bats and other animals.

The beauty of the area is evident from its manicure of glacier, tropical seas, mangrove swamps, freshwater swamps, lowland jungles, and highland cloudy forests. Unfortunately however, this God-perfectly created paradise is said to be in danger of extinction from natural resources exploitation including logging-induced deforestation. For instance, in preparation for the 2008 Summer Olympics which it hosted, the Chinese government placed an order of US$1 billion or 800,000 cubic meters of rainforest timbers for its construction works for the Olympic

---

[13] Les Roberts *et al, Mortality in the DRC: Results from a Nationwide Survey* (New York: International Rescue Committee, April 2003)

game. Also, the rate of forest conversion for planting of agriculture crops such as oil palm, smallholder agricultural conversion; and water pollution from oil exploitation and most especially by a Louisiana-USA mining corporation, Freeport McMoran and its mining partner, Rio Tinto through their mining operations are said to be threatening this serene environment into extinction.

Freeport McMoran's mine lies on Indonesian-controlled Irian Jaya, the western half of the island of New Guinea. It could probably pass muster as the most controversial mining operation in the world. Allegations of major environmental damage and human rights abuses have provoked investigations and protests in the US and riots on the island.

The Indonesian government maintained a 9 per cent share in the mine, enough to earn several hundreds of millions of dollars a year in royalties, taxes, and benefits, making Freeport Papua New Guinea's largest single taxpayer.

In the process of digging up vast quantities of ore-bearing rocks from the world's largest gold mine and third-largest copper mine, Freeport McMoRan has almost flattened the highest peak between the Himalayas and the Andes Mountains, damaging both the environment of Irian Jaya, and the lives of the people that live there. Water contamination is alleged to have caused skin rashes, stomach problems, bloody coughs, and deaths. Death has also been the fate of the fish and the sago palm trees that sustain the indigenous Komoro people. Among the Amunge people, the leaves of the vegetables and the skin of the pigs on which they live on have been discoloured from the pollution.

According to a Dutch study of the area around Freeport's mine, 20 - 40 kilometers of the Ajkwa river is so polluted with acid mine drainage and toxic waste metals that it will be hazardous to fishes and humans for 15 years. The study also reported that the river's flood plains estuaries and delta will be poisoned for 35 years.

Freeport McMoRan's operations is alleged to have caused major damage on the flourishing rain forests by dumping its tailings (the by-product of Freeport's mining operations which are a mixture of fine clays, flotation tailings, chemical precipitates, and slimes) thereby creating an artificial flood plain and killing the rain forest.

The extractive activities of Freeport McMoRan in Irian Jaya are believed to be the primary cause of damage to the rivers and rain forests of Irian Jaya. According to reports, Freeport McMoRan dumps one hundred and twenty thousand tons of tailings into the rivers each day and even at a point requested that the Indonesian government allows it to dump three hundred thousand tons per day which the Indonesian government approved without any difficulty.

Irian Jaya which is housing the world largest gold mine and the third largest copper mine is believed in many quarters to be one of the most environmentally exploited large islands in the world. The mine which is in the West Papua's Jayawijaya district is situated on 16,500-foot high, snow-capped Mountain. The Mine was at a point described as "having the world's worst record of human rights violations and environmental destruction."

The Freeport McMoran 78,000 tons of ore yield a day operation uses Lake Wanagon, an alpine lake considered sacred by the indigenous Amunge people, to dispose its waste rock as well as additional overburden, nearly all of which is dumped as mine waste and tailings into the rivers surrounding the mine. It is estimated that about 600,000 tons of rocks are dumped on daily basis into the rivers. The water from Lake Wanagon flows into the Ajkwa River system that flows down to the Arafura Sea. In addition, the mine dumps about 300,000 tons of waste tailings into the Ajkwa River every day. This dumping makes the water toxic and thick with silt, smothering and killing all plant life along the previously fertile river banks, in addition to the contamination of the source of drinking-water. The Ajkwa River was so badly polluted from the mine that Kwamki-lama residents were warned by Freeport McMoRan's own employees not to drink the water or eat plants that grow near the water.

The mine tailings that are dumped into a tributary of the Ajkwa River flow down steep mountain sides into rain forests at lower elevations, producing a desolate landscape, dead and dying trees with their broken branches protruding from tracts of accumulated grey sludge which is estimated to be capable of destroying 51 square miles of rain forest before 2040. It has also been estimated that about 3 billion tons of rock would have been processed by the time the mine is exhausted about 2040.

Reports also have it that sediments in the Ajkwa River have thirty eight times the amount of copper than the minimum amount required for calling an area contaminated. Further, water in Irian Jaya is so polluted with tailings and chemicals that the Indonesian government recommends that people in the area should no longer drink it. The extractive activities of Freeport have been fingered as causing massive pollution of the Ajkwa River and the disruption of the indigenous communities' subsistence agriculture which can no longer sustain crop production because their farm lands are polluted.

The level of destruction of the people's source of livelihoods attributed to Freeport McMoran is enormous, causing the missionaries to move the indigenous communities to the coastal areas where many of them perished from malaria attack.

When Freeport McMoRan executives and the tribal chiefs made their original deal, they had to use the interpreters which Anthropologists have suggested was very likely that there were misunderstandings between the chiefs and the Freeport McMoRan executives. According to the tribal chiefs, Freeport McMoRan's executives did not explain to them that the company was going to blast out a large portion of the mountain and remain in Irian Jaya for thirty years. But once the Amunge people realized that Freeport McMoRan was not leaving so soon, their chiefs had a confrontation with Freeport McMoRan executives. Within a month of this clash the Indonesian government ordered two thousand Amunge people to leave their homes. As the citizens were forced out of their homes, Freeport McMoRan and other companies settled in; first in base camps and eventually in small towns without consulting the indigenous populations. This was believed to have started a pattern of such forced expulsions and relocations.

Again, Freeport McMoRan's contract with the Indonesian government obligates it to give some supplies, provide housing, food, and transportation to the Indonesian military in exchange for their services in guarding the Grasburg mine.

Human rights abuses on the host and impacted communities in the extractive jurisdiction including incidents of beatings, stabbings, torture, and murder that occurred in December 1994, have been blamed on Indonesian army officials guarding the mine. Freeport McMoRan denied that any of its officials were involved.

Freeport is also alleged to be closely involved with the Indonesian army and their acts of violence dating back to a long history with the repressive President Suharto. Freeport McMoRan was the first foreign company to get permission to operate in Indonesia after Suharto took power in 1967. The company enjoyed many privileges in Indonesia as it is Indonesia's largest corporate taxpayer and provides the knowledge and skills to mine the massive amounts of ore from the Grasburg mines.

The Australian Council for Overseas Aid and the Catholic Church of Jayapura accused Freeport McMoRan of being well aware of the Indonesian army's abuses perpetrated against the mining host and impacted communities as Freeport McMoRan turned a blind eye while the Indonesian military killed and tortured dozens of native people in the area surrounding the mining concession. "Villagers were said to have been beaten with rattan, sticks, and rifle butts, and kicked with boots," one tribal leader told Catholic Church officials. "Some were tortured until they died." For instance, when the military in 1977, put down a native revolt, Freeport McMoRan was alleged to have contributed one million United States dollars (US$1M) towards this military operation, which resulted in the death of thousands of citizens.

When the Amunge people protest abuses by Freeport McMoRan, security officers and military personnel arrest them, beat them, accuse them of being members of separatist movement, and lock them up in small containers. Some of those arrested were said have met their death during continuous torturing session. In 1997, Yapenes Rony Magal came home after being beaten harshly and he died the next day in the hospital. However, before he died, he recounted to three eyewitnesses that Freeport McMoRan security had beaten him for eating without permission.

Freeport McMoRan has also been accused of causing conflicts between different tribes of West Papuan natives. Freeport McMoRan operates on ancestral lands of the Amunge and Kamoro tribes and the company has been accused of only listening to requests from these two tribes, and not the other project impacted people, thereby creating jealousy among the other impacted tribes. Also Freeport McMoRan's expansions of its mines operations has heightened tribal conflict as traditionally warring tribes are brought physically closer than they already need to be. No matter which tribe they come from, these natives have many of the same complaints against Freeport McMoRan and one of their biggest grievances is the social gap between them and the Freeport McMoRan workers.

In addition, they complain that foreigners more often get the better jobs in the company while the natives usually get the lower paying jobs and these natives comprise of only 15 per cent of Freeport McMoRan's workforce. Besides, natives are not allowed to enter Freeport McMoRan's stores or other public places in the Freeport city near the mine.

Freeport McMoRan was also accused of failing to pay mining royalties (or any other compensation) to the roughly 4,000 Amunge indigenous people displaced by the growing mine's concession area of 9,266 square miles since the company started its strip mining operation there in 1972. Many of the displaced people moved to the lowlands, where several hundred of them died from malaria and other diseases.

During March, 1997, several thousand villagers rioted in the towns of Timika and Tembagapura, located near the mine, blowing up one of Freeport McMoRan's ore pipelines. The reaction of the Indonesian military was swift and gruesomely morbid as they responded by sending United States-supplied OV-10 Bronco attack jets to strafe villagers. This action was code-named Operation Tumpas ("annihilation"). The Papuans claimed that thousands of men, women, and children were killed in this operation but the government admits to 900 deaths. Reports of the use of these counterinsurgency aircraft for this operation was suppressed and did not appear in the world press until a year later.

Even as Freeport McMoRan adamantly denied responsibility for alleged human-rights violations, it was alleged that the company and the Indonesian military responded to local indigenous protests by spending US$35 million to assemble barracks and other facilities to house and support 6,000 troops, "more than one soldier for each adult Amunge."

During May, 2000, the Grasberg mine's waste-rock disposal dam collapsed, killing four workers, and, according to one account, sending several 40-meter-high "tidal waves" of waste roaring down the Wanagon river towards Banti village destroying pig sties, vegetable gardens, and a burial ground' about seven miles downstream of the basin.

Within days of the spill, on May 8 and on May 18, about 600 Amunge people from Banti, Tsinga, and Arwanop started to protest against Freeport McMoRan. They blockaded the Freeport mine's access road, preventing workers' buses from entering the mine, shut down the company's offices in Jakarta and prevented about 1,000 Freeport McMoRan employees from entering their workplaces. In addition to protesting the environmental devastation and deaths caused by the spill, the protesters demanded that Freeport McMoRan Indonesia provide a larger proportion of its earnings to support local people in the impoverished province surrounding the mine. Roughly 100 police confronted the blockade but failed to break it until the Amunge representatives met personally with Freeport McMoRan's General Manager, Hermani Soeprapto, and presented their grievances against the company.

Following this spillage and protest, the Indonesian environmental officials informed Freeport McMoRan that the company must submit a comprehensive new plan and obtain government approval before opening a replacement dump for its waste rock. Freeport McMoRan also was instructed to clean up all destruction and pollution caused by the waste released during the accident. In addition, Freeport McMoRan was told to allow a criminal investigation by the police and government officials into the four men's deaths caused by the collapse. The company also was ordered by the government to compensate losses suffered by residents of Banti.

The dam at Lake Wanagon has failed three times (June 20, 1998, March 20, 2000, as well as May, 2000) due to the company's dumping of overburden. After the third breach, dumping was halted pending an investigation. The investigation, conducted by Freeport McMoRan and the Institute of Technology of Bandung (Indonesia) cleared the company to continue operations in January, 2001.

The construction of a dormitory town at Tembagapura in association with Freeport McMoRan Indonesia's mining operation at Mount Carstenz led to eviction of indigenous Amunge people, who were barred from entering the town, which houses as many as 20,000 Freeport McMoRan workers and family members. Freeport McMoRan moved and relocated the 1,000 inhabitants of a village, Lower-Waa, to the coastal lowlands, where, in one month, 88 of them allegedly died of malaria.

Indigenous peoples living in the area where Freeport McMoRan operate its enormous mine have no legal title to their lands under Indonesian law. Their land is classified as tanah negara (state-owned land) under the terms of the Indonesian Constitution. The same central government has granted Freeport McMoRan a legal right to use the land largely as it sees fit, with only the lightest of environmental oversight.

During October, 1995, after a lengthy investigation, the Overseas Private Investment Corporation (O.P.I.C.), a political risk insurance agency owned by the US government that provide insurance coverage for American companies doing business overseas, cancelled Freeport McMoRan's US$100 million political-risk insurance policy, citing environmental problems at the mine. Through a letter dated October 10, 1995, O.P.I.C. told Freeport McMoRan that the mine had "created and continues to pose unreasonable or major environmental, health, or safety hazards with respect to the rivers that are being impacted by the tailings, the surrounding terrestrial ecosystem, and the local inhabitants."

Freeport brought in Henry Kissinger to lobby the State Department to stop the policy cancellation (Kissinger's consulting firm was alleged to have received US$600,000 from Freeport in 1994), and Indonesian President Suharto made a personal appeal to President Clinton on behalf of Freeport McMoRan. The insurance policy was them reinstated.

Freeport McMoRan's Grasberg mine is not only the best-known and largest of several mineral-extraction projects which have stirred protests by the island's indigenous peoples. In Papua New Guinea, Rio Tinto's Panguna Mine, which was one of the world's largest open-pit copper mines before it was closed, dumped more than one billion tons of mine waste into the Pangana, Jaba and Kawerong Rivers, killing all aquatic life in the 480 kilometer river system. The waste formed a deposit approximately 20 kilometers long, as much as a kilometer wide and several meters deep along these rivers, with a copper-contaminated outwash fan in Empress Augusta Bay covering roughly nine square kilometers.

Rio Tinto, a mining partner of Freeport McMoRan in West Papua, is the parent company of C.R.A., an Australian mining company which operated the huge Bougainville Copper Mine, which was established by Rio Tinto during the 1970s while Papua New Guinea was still an Australian protectorate. A guerrilla movement started campaigning since the 1980s for compensation of Bougainville's traditional landowners, who have been dispossessed by the company's operations. An estimated 5,000 civilians were said to have been killed in the Bougainville area during the 1990s.

The Kutubu project, operated by the U.S.-based Chevron Oil Company, was Papua New Guinea's first oil-extraction project; critics alleged that its environmental and social impacts were much greater than reported, "including significant impacts on biodiversity, and the risk of [oil] spills, with the benefits to landowners being relatively low. Oil is piped 176 kilometers to the Kikori estuary. In 1996, non-essential staff had to be evacuated after threats from landowners dissatisfied with royalty payments.

The Porgera gold mine, maintained by Placer Pacific of Australia, (majority shareholder Placer Dome of Canada) has been accused of being one of the dirtiest operations on Papua New Guinea. It allegedly dumps 40,000 tons of tailings and waste rock daily into the Strickland/Fly River catchment basin. Environmental sampling has indicated levels of metals as much as 3,000 times the levels permitted by government regulations, which are not enforced.

Some of the world's largest transnational extractive industries have been active in exploiting the island of Irian Jaya for oil and solid minerals. Their roll call include Union Oil, AMOCO, AGIP, CONOCO, Phillips, ESSO, Texaco, Mobil, Shell, Petromer Trend Exploration, Atlantic Richfield, Sun Oil and Freeport McMoRan (United States of America); Oppenheimer (South Africa); Total (France); Ingold (Canada); Marathon Oil, Kepala Burung (United Kingdom); Dominion Mining, Aneka Tambang, B.H.P., Cudgen R.Z., and C.R.A. (Australia).

Mining concessions in the Ertsberg and Grasberg mountains, the Paniai and Wissel Lakes region, Fak Fak, the Baliem Valley, the "Bird's Head" western tip and the Papua New Guinea border area, have resulted in the dislocation and suppression of Papuan peoples which have sparked popular uprisings followed by military reprisals.

OK TEDI GOLD MINE

The Ok Tedi Gold Mine owned and operated by the B.H.P. Billiton of Australia started operation in 1981 at Ok Tedi in the western province of Papua New Guinea (PNG). This undertaking has been a subject of controversy because its tailings killed all aquatic life along 70 kilometers of the Ok Tedi River of Papua New Guinea following pollution by tailings and waste rock.

The members of communities affected by the mine's pollution blockaded and shut down the mine on November 25, 2001, costing the company close to US $1 million in lost production. Law suits brought by the indigenous landowners forced the company to reach an out-of-court settlement for $550 million Australian, with the communities.

Late in 2001, B.H.P. Billiton convinced the Papua New Guinea government to endorse three acts of legislation affecting the Ok Tedi Mine which would have provided the mining company the "apian way" to escape responsibility for the damage it has caused to the environment and communities living near the Ok Tedi in the western province of Papua New Guinea. If approved and signed, the agreements would have given B.H.P. and Ok Tedi Mining "unrestricted legal indemnity for the pollution and destruction caused now and into the future by the operations of the Ok Tedi mine. The mine's owners will have no obligation to stop tailings entering the river system in future, and will be permitted to increase the amount of copper it is currently permitted to dump into the river system." The agreements releases B.H.P. Billiton from any liability and the suit in Australia's Victorian Supreme Court; under the same agreements, landowners would also lose their common-law rights to enforce a 1996 settlement as well as any future legal rights to sue the mine for any environmental damage.

At about the same time, representatives from four indigenous communities living near the mine presented a petition demanding compensation and a share of B.H.P. Billiton's 52 per cent stake in Ok Tedi mine. The local people also demanded compensation for environmental damage from the date the mine opened 1981. However, the government of Papua New Guinea denied the request from local landowners to grant them 12 per cent of the benefits from the Program Trust Company, to which B.H.P. Billiton's interest in the mine was sold in 2002.

The then PNG's Prime Minister, Sir Mekere Morauta, called the Ok Tedi Mine a national asset stating that the closure of the mine could devastate the nation's economy and causes the ruin of communities that depend on it. This could be true given the fact that the mine accounted for 10 per cent of Papua New Guinea's gross national product and 20 per cent of its export income.

As the mine's owners were sealing their deal with the Papua New Guinea government, the Australian Conservation Foundation issued a report saying that nearly 70 kilometers of the Ok Tedi River has become "almost biologically dead," and 130 kilometers of riverbank have been "severely degraded." Fish stocks have declined between 50 per cent and 80 per cent, according to the mine's own internal report. Roughly 30,000 downstream landowners have lost their ability to live off their own land. A scientific Peer Review Group employed by the mine's management identified potential for a total collapse of the fishery.

By early 2002, after 21 years of massive operations at the Ok Tedi gold mine, B.H.P. Billiton officially exited Papua New Guinea, leaving behind 30,000 people displaced by environmental destruction and human rights violations meted against host Ok Tedi and Fly River people. The company on its own admitted in 1999 that the "sediment load" of waste rock or tailings has killed 90 per cent of the fish in the lower Ok Tedi River.

EUROGOLD EXTRACTIVE ACTIVITIES IN TURKEY

Despite the fact that the level of environmental, health and human rights abuses and damages were not as pronounced and also as negative as in other developing countries where extractive companies operate, the country of Turkey is not left out in the roll call on agitation for balancing development (sustainable development) in the extractive sector management

In the 1990s Turkey witnessed the most effective and visible environmental social movement against a multinational mining corporation, Eurogold, investing its resources to establish the first modern gold mine in the country. This environmental social movement was known as the Bergama Resistance.

The genesis of Bergama Resistance against Eurogold in Turkey can be traced to the 1985 enacted mining law that promoted and encouraged the involvement of foreign corporations in extracting national underground resources which was previously an exclusive reserve of only state-owned enterprises. The purpose of the new law was to massively open the country to increased foreign direct investment which was needed to help in the structural transformation of Turkey economy from its agrarian- based economy into an export-oriented economy.

One of the provisions of the Turkish 1985 Mining Law is the requirement that a prospective mining company in Turkey shall establish a company in Turkey in order to be able to operate a mining undertaking. Following the enactment of the new mining law, the Normandy Mining Corporation from Australia with its several other partners tied to this project including La Source from France and Inmet from Canada decided in 1989 to establish the company named 'Eurogold'. Its plan of action was to operate the mine for 8 years focusing on the extraction of 24 tons of gold and 24 tons of silver using a combination of open-pit and underground mining techniques, and processing was also going to take place on site through the process of cyanide leaching. Further, a tailings pond would be constructed.

To achieve its goals, the company required some land which was acquired through sell of lands by local peasants, especially Ovacık residents, as well as receiving permits from relevant government authorities to operate on state-owned lands.

First, the peasants' initial response to Eurogold was very much positive, shaped by lucrative land sales and hopes for well-paying jobs resulting from mine development.

Though there was very strong opposition to the mine through peasant activists in the movement from surrounding villages, the movement has come to be identified with Bergama, home to the ancient Greek settlement of Pergamum, which houses numerous historical sites such as the Acropolis which attracts tens of thousands of tourists every year; remnants of the Great Altar of Pergamum also known in Turkey as the Altar of Zeus.

Bergama though a city of tourism is a place also blessed with some of the most fertile lands renowned for high quality agricultural produce in the country.

Though the Bergama Resistance was sparked by the proposed Eurogold mine development, it was an itinerant yet a calculated reaction to the broad and far-reaching political economic changes being implemented in Turkey since the end of Turkey's last military government (1980-83), which brought with it a spate of political, economic and social transformations that can be verily referred to as 'neoliberal economicization' of Turkey.

The tourist town of Bergama which brought together the relatively prosperous peasants and a group of determined activists became a landmark of the turning point in environmental politics that transformed Bergama to the status of the hub of activism against Eurogold mining development in Turkey.

The Bergama campaign was reacting to the fear of cyanide leaching from Eurogold mine which was located in the village of Ovacık situated some12km west of the city of Bergama with a population of approximately 50,000 in the province of İzmir transforming to the epicenter of the demonstration.

The union so formed was encouraged primarily by the perceived environmental and public health risks anticipated from pollution by cyanide leaching process in the Eurogold mine. This peasant activists' campaign that was mobilized against Eurogold sparked a national discuss over the environmental costs of rapid economic growth through extractive companies activities in Turkey.

In 2002, Normandy was acquired by Newmont Corporation from the United States. In 2004, having failed to secure a permanent permit for extraction, Newmont decided to give up on the Ovacık mine and sold it to Frontier Pacific from Canada. At the end of another year of failed attempts to overcome legal and bureaucratic hurdles, Frontier Pacific too pulled out of Turkey by selling the mine to Koza Gold, which was an indigenous corporation from Turkey and a subsidiary of the Koza-Ipek holding corporation.

During the peak of the campaign against Eurogold, 17 villages in total came together to speak as one. The movement to resist the plans of Eurogold began with urban intellectuals and activists who entered into a dialogue with peasants from the villages surrounding the mine but unfortunately the population of Ovacık later largely moved to the side of the mine.

As the campaign continued to simmer, a number of key events and facts that gradually soured the existing support of the rural communities' relationship with Eurogold became a source of concern to the peasant activists. For instance, these villages depend on irrigated agriculture for their livelihoods and due to the nature of their crops (particularly but not limited to cotton) also attracted large numbers of seasonal workers both from the immediate area and farther out from Anatolia.

Also, some of the villages were distinguished by a number of social, geographic and historical characteristics. For instance, Ovacık had a large concentration of residents who had migrated from the Turkic communities of the Balkans and were resettled in the region on land that was considered by many to be relatively poor in quality.

Secondly, several of the most important villages in the movement—for instance, Narlıca and Pınarköy—were predominantly populated by peasants from the Alevi faith – a sect of Shi'ite Muslims who have historically suffered discrimination both during the Ottoman times and the modern Republic of Turkey and who are also renowned for their collective social democratic posture that lends itself well to political mobilization.

Further, explosions at the mine site used for exploration drilling damaged several nearby villages and infrastructure. Furthermore, a local woman blamed her miscarriage on the explosions and some peasants complained that their water supply was contaminated by chemicals used in exploratory drilling.

In addition, there was this fear that spread in the villages that the cyanide leaching process would pose grave and lasting danger to them if it were to contaminate the local water supply.

Despite the fact that Eurogold had commissioned an Environmental Impact Assessment (EIA) report which suggested that the proposed mine operation would conform to the highest standards and pose no environmental risks to the surrounding communities, there was this palpable fear by the peasants that the tailings pond that was built to contain the discharged effluents coming from the processing of the ore would pose a permanent risk since the region is characterized as a major earthquake zone.

This fear was advanced and given impetus by the strong reference to the earthquake that devastated the region in 1939 and the devastating earthquake of 1999 which helped push Turkey into a deep economic recession.

The tension generated by this Bergama Resistance case between the advocates of environmental protection on one hand and advocates of development in a rapidly developing and globalizing Turkey, involved a vast array of stakeholders and had moved from being a small peasant movement against locally unwanted company to a national case.

Several key actors in Bergama and İzmir became instrumental in channelling scientific knowledge on cyanide leaching and assisting in the creation of a coherent and sustained resistant movement. These actors include Sefa Taşkın, who was the mayor of the city of Bergama during much of the 1990s; who once launched an international campaign to repatriate the Great Altar back to Bergama with his extensive contacts within German civil society which gave him valuable experience in transnational activism needed in the Bergama resistance. His pedigree as a young, ambitious and patriotic national political leader with a bright future helped turned the seething anger of the villagers surrounding the Eurogold mine to him for a leadership role in the movement.

Birsel Lemke, who went on to win the prestigious Right Livelihood Award in the year 2000 was very instrumental in the formation of a coalition of actors especially during the early phases of mobilization and bringing her extensive connections in Germany to help convey significant support from international networks, for instance from Food First Information and Action Network (FIAN).

Senih Özay, is a lawyer based in İzmir, and with support of his colleagues from the İzmir Bar Association, provided enormous, vital and sustained legal support to the peasants in both engaging in effective acts of civil disobedience that stayed on the right side of the law and launching legal challenges against the operation of the mine.

Oktay Konyar, a vocal political activist and real estate broker, who was spectacular in organising key numerous colourful, high-impact demonstrations also performed the role of spokesperson for the peasant activists.

Friedhelm Korte, a German professor of ecological chemistry played a highly influential role in shaping the movement's focus on cyanide leaching as the key reason for opposing Eurogold.

Further, the movement drew groups of a variety of actors into the campaign by forging links with a number of NGOs and other emerging campaigns in Turkey, particularly the movement against the proposed Akkuyu nuclear power plant and by establishing mutually supportive relationships with other nationalist movements such as the one against the privatization of the Turkish Airlines that created more impact beyond Bergama.

The Bergama Resistance deployed various strategies to drive home their points. They embarked on such actions as blockading roads, driving their tractors into the mine to stifle operations, marching on the streets almost nude. They also used both national and international legal system to hone in their points and prevent the operation of the mine.

At the initial stage, the movement's legal challenges were mainly against Eurogold or its operational permits and the bureaucratic instruments of the Turkish state to the extent that the state technical branches of government were involved.

As the campaign legal process progressed through various stages of appeals, the Eurogold issue became deeply politicized moving up in the late 1990s to the Council of the State which is the highest administrative court in the country. Two decisions characterized both the unprecedented success the peasant activists had in the courts and the difficulty in translating these decisions into concrete outcomes outside the courtroom.

The first case had started in local courts and went all the way to Council of State (Danıştay) which was the highest administrative court in the country that had the final say in the matter. In 1997, invoking for the first time Article 56 of the Turkish constitution which states that "Everyone has the right to live in a healthy, balanced environment", the court ruled in favor of the peasants. In its decision, the Court stated that the environmental and public health risks of cyanide leaching amounted to a breach of peasants' constitutional rights.

Even Prime Minister Bulent Ecevit became involved, supporting the cause of Eurogold at the behest of the Australian Prime Minister John Howard who visited Turkey in 2000 and immediately following this ruling, Prime Minister Ecevit, in a move clearly designed to find a legal opening to overcome the verdict of the highest administrative court of the nation, instructed The Scientific and Technological Research Council of Turkey (TÜBİTAK) to prepare an assessment of the risks of cyanide leaching. TÜBİTAK came back with a report which indicated that cyanide leaching posed "zero risks". The government argued that the TÜBİTAK report essentially neutralized the decision of the court and gave Eurogold the green light to operate.

This set off another round of legal challenges, which saw the Bergama movement take its case to the European Court of Human Rights (ECHR). In its 2004 ruling, the Court decided that the state of Turkey had indeed violated the activists' procedural rights and awarded EUR 3,000 to each of the 315 plaintiffs involved in the lawsuit. Nonetheless, the ECHR declined to support the activists in their calls to order Turkey to shut down the Eurogold mine.

The Bergama Resistance demonstrated the power of effective and highly motivated social action to alter the course of national debates over environment and development and became a turning point for environmental politics in Turkey. The Resistance is remembered as the largest and most effective civil society mobilization for environmental protection in Turkey that solidly influenced national environmental policies, informed and inspired numerous other mobilizations that followed and popularized environmental politics at the national scale.

So far, Koza which is an indigenous corporation has become very much more effective than its multinational counterparts in navigating the labyrinth of the Turkish political and economic processes and has started to produce gold at the mine.

Did Koza succeed because it is an indigenous corporation willing to protect the interest and rights of Turkey people because the corporation understands the political arena it is operating?

Would a foreign company have acted the way Koza did to prevent mass human rights abuse and its well-known consequences in Turkey given their antecedents in other extractive jurisdictions?

These are just two foods for thoughts that need further research.

THE CHAÑARAL AND HUASCO IN CHILE
CHAÑARAL IN CHILE

Chañaral, which is the capital of the Chañaral Province in the Atacama Region of Chile is a small coastal town and was a major fishing center in the 19th century and later became a commercial port for the export of iron, copper and related products from the nearby mines.

In 1824, Diego de Almeyda discovered large quantity of natural copper deposits in the area. The discovery led to the founding of the town on 26 October 1833 as Chañaral de las Ánimas ("Chañaral of the Souls"). A few years later, Pedro Lujan discovered mineral ore at El Salado, where a mine was built, and in 1836 a shipping port was constructed. This was the reason Chañaral became the first mining industry in Chile to export copper. However, the great boom in Chañaral did not start until 1860, when A. Edwards & Company was inaugurated.

The copper extracted from the Chilean mountains has been estimated at 38 per cent of the world's production, making it the country's first economic resource. The copper extraction process requires complex physical and chemical treatments, and a layer of rock is considered rich if it is constituted of 1.8 per cent of the precious metal or more. The toxic sludge resulting from the extraction process contains high levels of copper, zinc, lead, mercury, molybdenum that is more than 21 contaminants, most of which are heavy metals, with concentrations reaching levels considered 20 to 50 times above what the human body can tolerate.

For a period of 37 years, (1938 to 1975) the Potrerillos and later (from 1959), the Salvador copper mines disposed of their mining tailings in the normally dry Rio (River) Salado which empties into the Chañaral Bay. It is believed that the disposal system introduced about 250 metric tons of tailings into the Chañaral Bay, forming a deposit up to 10-15 metre thick and creating an artificial beach above the high water mark and is permanently dry and extended approximately 1km by 4.5 km long and also the tailings extended out from the shoreline under the sea for a further 2 km or more.

Throughout this period, the River Salado relentlessly moved dangerous waste particles downstream from their origin located 150 kilometers upstream, in the mountains in the east, where copper is intensely mined. The town is slowly and silently dying, poisoned by the copper industry which *started in 1938, when the retention ponds used to capture the mining tailings from the Porterillos mine operated by the American company, Andes Copper Mining, started overflowing.* When the wind blows, the dangerous particles start spreading in the air that the people of Chañaral breathe all day long.

With the high level of marine and air pollution, as early as the 40's, Chañaral residents started noticing scarcity of seafood and fish and part of the population who went swimming in the ocean complained about the increase in skin deceases. In the end, the sediments built up into what is today the gigantic toxic beach. Despite the huge public outcry about the high level of contamination, the major concern of Andes Copper's management was over the silting of the Bay of Chañaral, where their export harbour was located. They were concerned that the build-up of the sand may impede their successful operations that guaranteed that millions of tons of copper were shipped around the world.

The disposal of mines tailings from the Potrerillos and Salvador mines into the Rio Salado, and their subsequent accumulation in Chañaral Bay, has been described by the United Nations Environmental Program (UNEP) as one of the Pacific's most serious cases of pollution.

Even, with the Nationalization of Andes Copper Mine by Chilean government and changing its name to CODELCO in 1971, nothing was done to reverse the water treatment problem in the Chañaral region.

Though it took another five years of continuous contamination but in 1975, the Chilean Supreme Court ordered the company to immediately put a stop to the pollution of the water, and to start decontaminating the area. Such a decision was the first in the nation's history, and instantly became symbolic for the communities that were fighting against unsustainable industrial development. It is important to recognize the fact that strong pressure coming from the international community and the engagement of the local population provided propulsion for this successful intervention.

The waters were very seriously polluted and the contamination levels kept building up. With the court order the company embarked on the construction of a concrete channel that year to take discharges from the Rio Salado, 10 km upstream from Chañaral beach, to deposit at Caleta Palitos" creek, located, between Chañaral Bay and the Pan de Azúcar National Park in an effort to move the issue further away instead of finding a permanent solution to the problem. This resulted in a further dumping of about 130 metric tons of untreated tailings at this location.

Several processes were then set in motion: CODELCO financed the construction of a concrete canal leading the wastes from the mine to a brand new tailing pond located north of the town of Diego de Almagro. According to the company in charge, the polluted sediments were separated from the water which was then released in the Rio Salado. Furthermore, a number of shrubs and bushes were planted in the vicinity of the contaminated sand dunes with the cooperation of the organisation in charge of the national park. The aim was to trap the toxic dust in form of a carbon sink when blown away by the wind, and therefore limit its dispersion. However a minuscule and pathetic total of less than two hectares of shrubs were planted, while the Chañaral toxic sand patch stretches over 500 hectares.

In comparison to the extent of the problem, these actions can be considered purely soporific, cosmetic and unsustainable. With a total cost of decontamination that could reach over an estimated 500 million United States dollars today, this situation is bound to worsen because neither the government nor the company is ready to make such huge investments.

The "Pan de Azucar National Park", which is principally famous in Chile for its exceptional biodiversity lies 30 kilometers north of Chañaral and it is the natural habitat for the Humboldt penguins, sea lions, dolphins and otters.

The activities of the mining companies reduced Chañaral to what was regarded as sanitary and ecological disaster; no doubt in 1983, the United Nations Development Program (UNDP) declares Chañaral as one of the most contaminated towns in the world; and the reality of the situation is truly dumbfounding.

In 1990, tailings disposal on the coast finally ended after 52 years and Copper Bay introduced an initiative to rehabilitate Chañaral Bay which received a high degree of support from all involved parties, specifically government, environmental groups and local communities. The project provided a tremendous opportunity to clean up the environment while at the same time rewarding its investors with capital growth.

In December 2003, the President of Chile at the time, Ricardo Lagos, in staged performance to prove that the water was clean and safe went for a swim in the bay of Chañaral; a marketing strategy to ensure CODELCO could get an environmental certification.

But the pressure on the sanitary consequences reached a critical point for the authorities. A health study of Atacama (Servicio de Salud de Atacama) proved that in the years between 1990 and 2007, the main causes of mortality in Chañaral were linked to tumours (24 per cent), circulatory deceases (21 per cent) and respiratory problems (13 per cent).

"In comparison, in France the mortality rate linked to tumours is of 2.4 for 1000 inhabitants (source: Cepi DC). In Chañaral, it is 128.3 for 1000 inhabitants"

Although these figures may seem significant, no study has yet linked the environment scandal in Chañaral to the health problems of its local community. However, a Doctor explains that *"international publications clearly demonstrate the direct health consequences related to heavy metal expositions."*

The major action being taken by CODELCO to remediate the environment is to pour blames on its predecessor Andes Copper Company".

The 2009 report by a group of deputies challenged the president of Chile at the time to evaluate the risks and the consequences for the wellbeing of the Chañaral community. The health minister was requested to intervene immediately in the matter to protect the most vulnerable, the young and the elderly. However, no actions were taken to follow up with the report.

Is the desertion of their hometown the only solution left to the inhabitants of Chañaral, as it was the case for the people of Potrerillos? Located right next to the Rio Salado, this little town was declared "*saturated by sulphur dioxide and particulate material*" by the government. The entire population, mostly constituted of miners and their family, was displaced to El Salvador, 25 kilometers away.

Travelling along the bay, there is now a road built where the ocean once was and Caleta Palitos, where a second toxic beach can be found, is situated less than 10 kilometers north.

The northern section of this area is part of the national park, where 25 000 tourists come to visit every year. The question is to know if one of the most important natural reserves of the country is being affected by the pollution. Without a doubt, the answer is yes.

In 2011, a veterinary student dissected a sea lion only to find that its bones were blue as a result of contamination from copper and it is said that "*Pan de Azucar's wildlife isn't also spared, as dead birds are regularly found.*

The government authorities have demonstrated clearly their lack of capacity to frontally confront the environmental problems, sending the Chañaral residents into despair and probably to suffer the same fate as the Cofanes, Siona-Secoya and the Tetetes of the Amazon Basin. The government however keeps telling the residents that the mine of El Salvador is an opportunity for wealth creation, but for the majority, it only brings poverty.

The Chañaral town that used to parade itself as a small fishing town with a bright future has lost hope because there are no more fishes and seafood to be found in abundance as they are now living in a completely sterile marine environment. By poisoning their air and water, the industry is slowly killing an entire Chilean region, silencing its population along the way.

THE HUASCO IN CHILE

The Pascua-Lama mine project located at the border between Chile and Argentina operated by Barrick Gold, a Toronto-based mining company is another case worth a mention.

The Pascua-Lama is estimated to hold "in its bowel", 17 million ounces of gold and 635 million ounces of silver deposits, with 75 per cent of the deposits located in Chile while 25 per cent is located in Argentina.

The mine was scheduled to start production in 2014 with estimated production of about 850,000 ounces of gold a year, and 30 million ounces of silver in the first five years using up to 27 tons of cyanide and 33 million litres of water per day to extract the gold.

However, Barrick's plans for the project met some challenges that caused the project plan to change over time. These include:

1. The proximity of the mine to the glaciers, the source of the Huasco River.
2. The proposal to "Transplant" the three glaciers in mountain, in order to gain access to the ore deposits underneath them, by moving the three glaciers to another glacier with which they were to bond has been scrapped. This is said to be equivalent to the removal of 20 hectares of ice, with an estimated volume of 300,000 to 800,000 cubic meters, and that will cause serious environmental harm.
3. The 18 indigenous Diaguita Huascoaltino communities living in Chile's Huasco Valley, claimed ownership to the land where Barrick's airstrip in Pascua Lama, is built on sacred land without consultation as provided for in ILO Convention 169.
4. The fear of contamination of the river source which the 70,000 farmers in the Huasco valley depended for their livelihoods sustenance.
5. The Comunidad Agrícola Diaguita Los Huascoaltinos (the Diaguita Huascoaltinos Indigenous and Agricultural Community) in Chile opposes the Pascua Lama mine because "Barrick Gold seeks to extend the Pascua Lama project to the top of the Pachuy Ravine, which is located within the Community lands recognized by the 1997 domain title, however, the Mining Code requires the Mining corporation to take over the community ancestral lands.
6. Although the mining work has not fully begun, however, there have been roads built by the mining company, and the exploration activities carried out in the high mountains have created severe wetlands and large-scale landscape deterioration affecting the river drainage capacity.
7. The fear that some residents will be forced to relocate away from their homes the same way as their neighboring town of Copiapó located 140km to the North, which for 10 years received no rainfall causing the Copiapó River to dry up. This allegedly resulted from mining and agricultural activities in the region; making the future of Copiapó residents unknown as they struggle with the concept of leaving their homes and their land due to water scarcity.
8. Opponents to the mine argued that the project will negatively affect the river flows and the farmers who are dependent on the water supply, as well as reduce the water quality.

These concerns caused public protest in Chile, including demonstrations and petitions presented to the Chilean government.

To ensure the project gets underway, the Chile and Argentina Presidents signed and adopted the *"Mining Integration and Complementation Treaty"* in 1997 which gave clearance to the investors to explore and exploit mineral deposits that span the border between the two countries. The bilateral treaty was ratified by their respective national legislatures in 2000.

The opponents of this project concluded that the treaty was unconstitutional.

Following all these controversies and subsequent "environmental review" carried out over more than two years the government authorities imposed 400 conditions on the company which the company must meet in order to operate the mine.

Barrick Gold contends that the project is environmentally friendly in terms of water treatment, and that the project will create 5,500 direct jobs during the mine's construction phase. It contends also that underground mining methods are not economically feasible for the mine, only open pit methods. It states further that its US$1.5 billion investment "would be directly invested in the Huasco province in Chile and San Juan province in Argentina", and that it has "identified more than 600 potential suppliers from Chile's Region III" in pursuance of its policy of sourcing local goods and services, and that through its corporate social responsibility framework "sustainable development projects have been and will continue to be a priority for funding to the tune of millions of dollars focused in the areas of education, health, infrastructure and agricultural improvement".

In June, President Piñera's government announced its strategy for promoting development in the northern Region of Arica and Parinacota. The Region of Arica and Parinacota is where the Aymara people currently live and where their ancestral lands are located. Northern Chile is also the region where Chile's mining enterprises are concentrated, and chief among Piñera's development proposal was the continued expansion of the mining industry. Most notably, he announced plans to open 40,000 hectares of the Parque Nacional Lauca to copper mining activities. Some of the regional government officials support the initiative, as it is predicted to bring in US$2 billion in investments, quadruple the region's GDP, and create 9,000 new jobs.

The Coordinadora Aymara de Defensa de los Recursos Naturales de Arica Parinacota issued a statement condemning Piñera's plans and summarizing their chief concerns, particularly in light of Chile's international human rights and biodiversity obligations under ILO Convention 169 and the Washington Convention, respectively.

Aymara communities and leaders, however, raised serious concerns about the government's plans. Studies carried out by the University of Chile indicates that the overwhelming majority of Parque Nacional Lauca is Aymara property, thus raising questions about the government's legal right to open the lands to mining. The Aymara additionally assert the government's obligations under ILO Convention 169 to consult with the affected communities. There are also concerns being raised about the environmental impacts of more mining activities, particularly on water resources.

On August 28 in Arica, the *Red por la Defensa del Medio Ambiente de Arica y Parinacota* (Network for Environmental Defense of Arica and Parinacota) organized a march in protest of national government development policies and programs that are expected to have a negative environmental impact in the region of Arica and Parinacota.

On November 14, the Coordinadora Aymara de Defensa de los Recursos Naturales (Aymara Coordination for Natural Resource Defense) issued a declaration concerning the proposed mining activities in northern Chile, and particularly the lack of consultation with Indigenous people.

In June 2011, the third chamber of the Chilean Supreme Court issued its ruling in a case which sought to halt the "Catanave mining project" in northern Chile on the basis that the government had not carried out consultation with the affected Aymara indigenous people. The Arica Court of Appeals had rejected the claims of the Aymara communities that filed the suit and the Supreme Court upheld that decision.

On Friday, March 30th, the Chilean Supreme Court decided their second case in a month where they ordered consultation with Indigenous people. The case came out of northern Chile where Aymara communities sought to stop prospective drilling occurring on their territories. Ultimately, the Court ordered the company, Compañía Paguanta S.A. to halt its actions until an environmental impact study is performed and the Aymara people are consulted as is required by ILO Convention 169.

On Wednesday, April 10th, 2013, the Court of Appeals in Copiapó, Chile unanimously ordered all work at the Pascua Lama mine halted until further notice. The order was a result of a lawsuit brought by the Diaguita communities of the Huasco Valley in northern Chile on the allegation that the mine's work was violating multiple environmental regulations including regulations that protect three glaciers in the area from environmental harm and contamination. The decision marks the second time in the past 12 months that a Chilean court has halted a mine based on a lawsuit brought by Indigenous communities. The mining company was also fined US$16 million and ordered to resolve dozens of environmental issues.

Before the April 10th, 2013, Court of Appeals in Copiapó, Chile, ruling in the case brought by the Diaguita communities of the Huasco Valley in northern Chile, in another similar case decided on Friday, April 27th, 2012, Chile's Supreme Court issued a decision in favor of indigenous peoples. This adds to the list of recent decisions that are slowly advancing in Chile the indigenous right to consultation. This latest decision involved the *Comunidad Agrícola los Huasco Altinos* and the "El Morro" mining project, owned by Socketed Contractual Minera El Morro. The decision will halt the large mining project until consultation with the community takes place.

In almost every developing country where Barrick undertakes its mining activities, there has always been one accusation or another against the company. For instance the company has to deal with the following issues:

- A court case in the British High Court seeking damages for the death and injury of local villagers in Tanzania.
- Communities in Papua New Guinea making urgent calls for resettlement away from the mine site and community members seeking compensation for killings and sexual assault.
- Communities in the Dominican Republic seeking urgent relocation away from the contaminated mine site, and to be compensated for their economic losses from dead cattle and contaminated produce.
- Indigenous Diaguita communities in Chile fighting to stop the Pascua Lama project, which does not have their consent and has been poisoning their scarce water resources.
- Communities on Marinduque Island in the Philippines seeking damages from a mine tailings disaster, considered the worst mining disaster in the Philippines.
- Barrick being sued for fraud by their own shareholders because the company management team allegedly lied about meeting environmental regulations in Chile and the cost and time estimates for their Pascua Lama project.

THE NIGER DELTA REGION OF NIGERIA

The Niger Delta region of Nigeria could be rightly referred to as the oil and natural gas capital of Nigeria. The original area described as the Niger Delta is the area of Nigeria that lies between the estuaries of the Benin River to the West and Cross River to the East of the River Niger itself. But today the area referred to as the Niger Delta comprises of the nine oil-producing states of Abia, Akwa-Ibom, Bayelsa, Cross River, Delta, Edo, Imo, Ondo and Rivers.

"The Niger Delta stretches about 435 kilometers (270 miles) along the Atlantic Coast and covers over 20,000 square kilometers. It also ranks among the world's largest wetlands and definitely, the largest in Africa". The region is crisscrossed by a web of creeks that link together the main rivers of Benin, Bonny, Brass, Cross, Forcados, Kwa-Ibo, Nun, and other rivulets and streams (all estuaries of the Great River Niger). "It's mangrove forest covers about 6,000 square kilometers with high biodiversity species of flora and fauna". It is a land endowed with human and natural resources. It is a "land flowing with milk and honey".

The Niger Delta has immense wetlands in the southern region of Nigeria where thousands of kilometers of waterways and creeks streak throughout communities where oil and gas development activities in Nigeria take place and where over 80% of Nigeria's financial resources are generated, yet many people in the area live on less than $2 a day, despite the riches that straddle under them.

The people of the Niger Delta include the Okrikas, the Ijaws, the Kalabaris, the Efiks, the Ibibios, the Urhobos, Ibos, Ilajes and the Itsekiris etc. They enjoyed exclusive rights over the waters in their area and used the rivers and the sea for their economic advancement–fishing, trade, and means of transportation. The land is another source of livelihoods as the Niger Delta people engaged in farming, hunting activities and harvesting of wild edible fruits and herbs and fuel-woods.

Oil prospecting in Nigeria dates back to 1908 and was started by the German company, Nigerian Bitumen Company, but the operation ended in 1914 with the commencement of the First World War. The British government on November 4, 1938 granted an exploration license to Shell Overseas Exploration Limited and D'Arcy Exploration Company Limited to jointly explore for oil throughout Nigeria but their exploration activities ended in 1940 because of the Second World War. However, after the Second World War in 1945, exploration activities resumed in 1946 with British Petroleum (BP) joining Shell Oil Company to form Shell BP that finally discovered oil at Oloibiri, in present day Bayelsa state in Niger Delta in 1956, leading to the first export of Nigeria crude oil on February 17, 1958. Since then, there have been many more discoveries of oil in the Niger Delta, leading to Nigeria being ranked the fourth oil-producing country among the OPEC countries and the biggest in Africa.

Ordinarily, the Niger Delta from where all the oil revenue and about 80 per cent of Nigeria's total revenue and 95 per cent of its foreign exchange earner should be a gigantic economic reservoir of national and international importance due to its rich endowments in crude oil and natural gas resources which feed methodically into the international economic system, in exchange for massive revenues that carry the promise of rapid socio-economic transformation within the delta itself. But in reality, the Niger Delta is a region warped and suffering from environmental pollution and human rights abuses, governmental neglect, crumbling social infrastructure and services, high unemployment, social deprivation, abject poverty, filth and squalor, and endemic conflict.

The Niger Delta has produced several hundred billion dollars' worth of oil since independence in 1960, but has derived appallingly little or nothing from all that wealth because of high level of corruption by both military and civilian leaders who have stolen much of the oil-wealth. The former Economic and Financial Crimes Commission (EFCC) boss, Mallam Nuhu Ribadu, indicated that as much as US$380 billion has been lost to corruption and waste between 1960 and 1999.

Human rights impacts of these losses are profound as funds that could have been spent on basic healthcare and basic education for citizens have instead been squandered or embezzled, while the Niger Delta public schools and health facilities have been left to crumble and wither away; the impact being decayed vital public services.

The ecology of the Niger Delta area has been severely bloodied, degraded and devastated from oil exploitation by the multinational oil companies operating there. In the Niger Delta, extractive industrial operations with few environmental controls have contributed greatly to habitat destruction. Oil operations in the Niger Delta region, for instance, have opened up new channels enabling salt water to intrude farther upstream. Regular oil spills contaminate the environment, and gas flaring contributes to air pollution. Water resources in the Niger Delta are also under increasing threats. Many extractive industries blatantly discharge their toxic effluents into waterways and use few controls to prevent modifying the landscape. With the large-scale removal of forest cover, erosion and resulting siltation continues to degrade waterways even further.

Community agitations for fairness and justice have been answered with repressive policies triggering resistance by the Niger Delta people through such activities as peaceful demonstration on the streets, riots, kidnappings, sabotage of oil pipelines, seizure of oil platforms and general harassment of oil workers, which started at a low-level cascading through outright resistance to open confrontations with the oil companies and Nigerian government. Militant groups have carried out widespread attacks on oil infrastructure, cutting more than a third of Nigeria's oil production, at their peak in 2006.

Royal Dutch Shell has been operating its extractive activities in the Niger Delta longer than any other oil company. This and other oil companies are alleged to have perpetrated and escalated violent community clashes in which entire communities have been destroyed, with billions of dollars in revenues lost by the government and the oil companies.

The Royal Dutch Shell has been severally accused of funding armed gangs in Nigeria which in no small measure had fueled human rights abuses in Africa's most populous nation. However, Royal Dutch Shell has vehemently denied the allegations.

In face of this vehement denial, Platform, a London-based non-governmental organization monitoring the oil and gas industry, came out with a 75-page report in which the Anglo-Dutch Oil Company was accused of paying government forces that attacked, tortured and killed Nigerians living in the Niger Delta creeks.

According to Shell, "We have long acknowledged that the legitimate payments we make to contractors, as well as the social investments we make in the Niger Delta region, may cause friction in and between communities. We nevertheless work hard to ensure a fair and equitable distribution of the benefits of our presence," Shell said in a statement in response to the report.

"In view of the high rate of criminal violence in the Niger Delta, the Federal Government, as majority owner of oil facilities, deploys Government Security Forces to protect people and assets. Suggestions in the report that SPDC (The Shell Petroleum Development Company) directs or controls military activities are therefore completely untrue."

However, the company said it would look into recommendations made in the Platform report.

In 2009 there was an agreed Amnesty Program between the government and the Niger Delta militants which led to thousands of militants in the Niger Delta laying down their weapons thereby reducing drastically major sabotage activities against oil infrastructures although there still exist some pockets of militant activities, community grievances and unrests.

Shell and other Multinational Oil Companies operating in the Niger Delta has persistently denied responsibilities to environmental damages caused in the Niger Delta through their exploitative activities but instead blame the majority of oil spills and subsequent environmental degradation on sabotage or oil theft by Niger Delta militants.

However, lack of tangible evidence to support its claims forced Shell to admit liability for oil spills in the Ogoni region of the Niger Delta.

The crises-torn oil-rich communities in the Niger Delta are no strangers to repression and brutal killings by government security forces in coalition with multinational oil companies, through first, the Rivers state Internal Security Force, comprising the Army, Navy, Mobile Police, Regular Police Force, armed supernumerary Police recruited and trained by the Nigerian Police Force but paid by the Oil Companies as well as armed private security guards; created in response to the Ogoni crises, to repress local residents of oil-producing states in the Niger Delta. The Rivers state Internal Security Force has now been replaced by the Joint Task Force (JTF), comprising the Air force, Army and the Navy.

Apart from using the armed forces to craft carnage in the Niger Delta, the government uses the Oil Pipeline Act of 1956, the Petroleum Act of 1969 and the Land Use Act of 1978, the Treason and treasonable offences Decree of 1993 to intimidate, repress, harass, maim, and murder Niger Delta people.

The Ogoni case is well documented internationally. This case involved the surreptitious murder in 1995 of nine Ogoni citizens under the leadership of Ken Saro-wiwa. These Ogoni indigenes were protesting continued abuse of their human rights and pollution of their environment by Royal Dutch Shell who was alleged to have sponsored another group of people from the same community to oppose the actions of the Movement for the Survival of Ogoni People (MOSOP) led by environmental activist and play-write, Ken Saro Wiwa. Royal Dutch Shell was alleged to be complicit in this murder in connivance with the late General Sanni Abacha led military government.

Apart from the Ogoni extrajudicial killings, the event of September 9 and 11, 1999, in which about 100 soldiers deployed from Elele Barracks in Rivers state and joined hands with the police with their marching orders to shoot on-sight saw to the destruction of the Black Market area of Yenagoa, in Bayelsa state, as they went from house-to-house in search of community residents to kill.

With this marching order, anyone found running was shot-on-sight and people who jumped into the river to escape alive were sprayed with live bullets. "A group of soldiers and police in violation of the Law they swore to protect, "the life and property of Nigerian citizens", jumped into three speed-boats, cornered the young boys trying to swim to safety, and sprayed the young boys with bullets and they went to meet their creator earlier than scheduled".

"About 100 *Ijaw* Youths were shot dead in the streets of Yenagoa and nearby communities, by the soldiers and police deployed for this mission. Dead bodies including that of a pregnant woman with a baby strapped to her back were seen floating in the Taylor creek on September 11, 1999."

Within few months after being sworn in as the civilian president in 1999, President Olusegun Obasanjo ordered a military assault on Odi community in Bayelsa state. The community was annihilated and it became a ghost town. According to Environmental Rights Action/Friends of the Earth Nigeria (ERA/FoEN), the march on Odi community claimed about 2,483 casualties, comprising 1,023 females and 1,460 males.

Once again the Niger Delta people for their contribution to the socio-economic development of Nigeria suffered huge casualties when the JTF comprising the Air force, Army and Navy, under the command of Major-General Sarkin Yakin Bello at the order of President Umaru Musa Yar'Adua, in May 2009, landed in the sacred Niger Delta communities in Delta state and acted with a mixture of pedantry and disbelieve to reconstruct the September 1999 cruel massacre of innocent Odi civilians and the black market area of Bayelsa.

The reason for their invasion according to the Nigerian Army Director of Defense Intelligence (DDI), Colonel Chris Jemitola, in an attempt to justify JTF offensive against the mostly Ijaw communities in Delta state said "a rescue team sent to rescue the ceased vessel *MV Spirit* and its crew members were ambushed and captured, tortured and killed and the contents of the ship stolen". However, the JTF Information Officer, Lieutenant Colonel (Lt-Col) Rabe Abubakar in his own differing account said "the militants of the Movement for the Emancipation of the Niger Delta (MEND), ambushed the troops during an escort of a vessel meant to deliver products to Chevron Oil Company and they killed a Lt-Col, a Major and five other ranks on May 15, 2009, for this reason the JTF sealed off Major entry and exit points on Warri waterways; a number of gunboats and helicopters deployed to comb the creeks. During the duration of this operation no boat would be allowed into that area unless with top military clearance.....Besides stifling supply to the militants, the measure is aimed at cutting basic supplies including food, water and medical aid to the militants who are injured".

An eye witness, His Royal Highness (HRH) Benenimbo, narrated his experience thus. "At about 10:45 this morning, as I was sitting very close to the waterside, I saw a naval warship, a JTF gunboat and their jet fighters moving towards Okerenkoko town in Warri South Local Government. Then my pastor and I had to run for our dear lives. They are still in my community, as I am texting you this message, destruction is going on there without anybody exchanging fire with them. The entire community is helpless. We should remember that one day God will bring our work to judgement".

The JTF information officer Lt-Col Rabe Abubakar actually confirmed this when he said: "in continuation of the search and rescue operations in some militants' hideouts, The JTF has moved into the outskirts of some communities where suspected militants are hiding and still holding expatriates kidnapped last week......we call on all of you to aid in the extradition of these miscreants for your safety and interests".

The JTF commander informed that "the military operations in Gbaramatu Kingdom were prosecuted using pinpoint surgical helicopter attack before the deployment of ground force". This operation according to the Army Chief "involved men of the Army, Air Force and the Navy". The people that witnessed the commando-styled massacre said, "The JTF who appeared and reappeared like serial killers applied their skilfulness with zombie-like precision" operating like whose first best option available was quick elimination". Deploying such assault weapons as Soviet Kalashinikov (AK 47), American M-16 rifles, Uzi machine guns, Rocket-propelled grenades (RPG), stinger missiles among others, to mow down unarmed civilians in the mostly Ijaw populated communities of Niger Delta.

To start the maneuver the military cordoned off Miller Waterside and conducted a house-to-house search where over 20 Ijaw youths were arrested and whisked away to face the spirits of the sun and river for their final prayers. The JTF then started the bombardment of women, children and the elderly. This invasion was described as the crux of obscene mission against defenseless innocent civilian women, children and the aged who found themselves in the target range of military professionals.

The JTF inflicted severe punishment on the Niger Delta communities creating widows and orphans, leaving massive property and human collateral damages, and "the smell that triggered an avalanche of buried memories" of Nigerian civil war, Liberia, Sierra Leone, Somalia, Sudan, Vietnam, Normandy, Kosovo, Cambodia, Beirut, Lebanon, Kuwait, Hiroshima, Nagasaki, Iran, Iraq etc. During the offensive every fixed place became truly unsafe for civilians who had to move, duck and hunker for their lives.

At the Nigerian Ports Authority (NPA) Warri, for instance, workers hunkered in their offices to avoid being hit by stray bullets. The JTF sped through NPA road firing guns, with reports of bullets hitting unarmed innocent bystanders and those running for their lives. One woman identified as mama Boloko was shot in the head, neck and thigh by the military personnel. As she lay agonising in her own pool of blood, some "good Samaritans" daring flying bullets rushed to rescue her and she was rushed to the central hospital.

At the central hospital junction near CAT Company, a stray bullet fired by the JTF hit a bus driver who along with other victims, were rushed to the central hospital in a Kombi Bus with registration number Lagos XS 06 APP. At Essi junction near the Nigerian National Petroleum Corporation (NNPC) Zonal Headquarters, people witnessed JTF soldiers dragging a bloodied youth into their waiting van for final escort into the waiting hands of the Almighty God.

The killings and maiming of innocent people including women, children and the aged in the Warri creeks have been described as genocide, dastardly, callous, inhuman, barbaric and insensitive; while the destruction and complete razing of Okerenkoko, Oporoza, Kunukunuma, Pertorukorigbene, Kurutie and many other communities were christened ethnic cleansing.

Equally, the government and multinational oil companies have been allegedly blamed of instigating the impoverished communities that lived peacefully together through the ages to turn against each other with sophisticated weapons that could only have been supplied by the military, in a "mutually assured destruction (MAD)", resulting in a lot of local communities in the Niger Delta to be sacked. Death has consumed thousands of innocent people in the course of instigated communal conflicts. Pipeline explosions have also consumed innocent souls making life caustic at best, brining pains, massive destruction and death of unqualified magnitude on the Niger Delta people.

Apart from the human rights issues reported above, the Niger Delta is on a daily basis confronted with environmental pollution from the extractive activities of the oil and gas companies operating in the area. The waters are polluted from oil spillage, killing the fishes and other marine mammals; the land is polluted from dumping of waste and oil spill and the air is polluted through gas flaring that drive away animals deeper into bushes and acid rain. Apart from these, there is a high level of poverty among the majority of the population.

ECUADOR

Ecuador is divided into three distinctive natural zones namely the Tropical Pacific Coast, the High Sierra, and the Amazon Basin, also known as the orient.

The story of the Huaorani people of the Amazon rainforest in Ecuador is not different from their counterparts in other jurisdictions where extractive companies operate.

Petroleum exploration in the Ecuador dates back to 1878 when Ecuadorean National Assembly decreed exclusive rights to M.G. Mier and Company to extract petroleum, kerosene, and other bituminous substances in the Santa Elena Peninsula. In 1937, Shell Oil Company received the first concession to produce petroleum in the Amazon Basin. Texaco was invited in 1964 along with Gulf Oil Company.

On arrival, Texaco and Petro Ecuador (the Ecuador's state oil company) formed a consortium to extract petroleum in the Amazon region. Petro Ecuador owned 62.5 per cent while Texaco owned 37.5 per cent share of the undertaking and was the operator of the joint venture project from 1964 to 1990 when it handed over the operation to Petro Ecuador, but still retained its 37.5 per cent of its capital investment. By 1992, the consortium contract ended and Texaco relinquished its share to Petro Ecuador and this arrangement gave the Ecuadorean oil company 100 per cent ownership of the operation.

During the period of 1964 to 1992, Texaco allegedly caused a lot of environmental damage and human rights abuse in its area of operation. The numerous environmental issues that resulted from Texaco's operation include the pollution of the rainforest and rivers of Ecuador and Peru. The company was said to have improperly dumped toxic effluents from its operation into the local rivers instead of pumping them back into the empty wells. Toxic materials were also allegedly burnt, dumped into landfills and even spread on the local dirt roads.

In fact, Texaco was accused of discharging 4.3 million gallons of toxic materials into the orient; 20 billion gallons of waste water discharge was recorded; opening up about 2.5 million acres of forestland to colonisation and road construction; and spilling of about 17 million gallons of crude oil into the orient. All these had direct negative consequences on the environment and the people of the orient.

On the human front, some indigenous people like the Tetetes went into extinction, while the population of others like the Cofanes and Siona-Secoya reduced drastically. The people suffered skin rashes, stomach pains and chronic headaches and fever from using polluted water. Some took to alcoholism while some females took to prostitution. The environmental pollution and human rights abuse visited on the Ecuadoreans led to Texaco being sued in Texas and New York[14].

BURMA (MYANMAR)

In 1988, the Burma military establishment seized power from the civilian government and this action prompted massive non-violent protest, led by the pro-democracy movement against the military, throughout the country. In an attempt to quash the protest and restore normalcy or something near to it in the country, the military elite created the "State Law and Order Restoration Council (SLORC)", and imposed Martial Law in Burma and renamed the country Myanmar.

---

[14] These commentaries were contained in the legal action brought against Texaco under the United States Racketeer Influence Corrupt Organization Act (RICO).

On May 27, 1990, SLORC held multiparty elections in which the opposition party, the National League for Democracy (NLD) party, led by Tin Oo and the 1991 Noble Peace Laureate, Aung San Suu Kui, won 82 per cent of the parliamentary seats. SLORC promptly arrested the NLD leaders and also intensified its campaign of repression throughout the country.

In July 1992, Total S.A. (Total) and the Myanmar Oil and Gas Company (MOGE), the state oil and gas company controlled by SLORC signed a production sharing contract for a joint venture gas drilling project in the Yadana natural gas field. In 1993, UNOCAL which is a US-based oil exploration company formally agreed to participate in the joint venture project whose objective was to obtain natural gas and crude oil from the Andaman Sea and transport them through pipelines across the Tenasserin region of Myanmar into Thailand.

The joint venture partnership agreement requires SLORC to act as agent of the joint venture operation to clear the forest, level the ground, provide labor, materials and security for the project, while UNOCAL and Total are to subsidize the activities of SLORC.

Having been given these responsibilities, SLORC must perform effectively so the joint venture partners will succeed in their operations. What SLORC did was to use its military intelligence unit and police force to violently intimidate and relocate whole villages, enslaved farmers living in the areas of the proposed pipeline routes, forced the farmers to clear the pipeline routes and build Headquarters for the pipeline construction workers as well as prepare military outposts and carry project supplies and equipment to work sites.

As a result of the forced relocation, many villagers lost their crops, and livestock. Also, due to the prevalence of SLORC'S forced labor practices, many farmers were unable to maintain their homes and farms, and were forced to flee. Woman and girls in the Tenasserin region left behind after the male family members have been taken away to perform forced labor, became targets of rape and other sexual abuses by SLORC officials. Some of the women were also subjected to forced labor. There were reported cases of gang-rapes by SLORC officials guarding the women during the period of forced labor. The villagers also reportedly suffered assaults, torture and extra-judicial executions, as well as other human right violations in furtherance of the Yadana gas pipeline project.

The US$1.2billion Yadana gas pipeline from Burma to Thailand, developed by the United States-based Unocal, the French company Total, the state-owned petroleum Authority of Thailand [PTT] and MOGE, generated massive human rights controversy.

The incidents of human rights violations occurring in Myanmar did not spare Thailand at all. Residents of both Myanmar and Thailand witnessed egregious violations of their rights. In Thailand, for example, the environmentalists protesting against the gas pipeline project were arbitrarily arrested.

All these acts were a clear violation of state law, federal law, and customary international law. These violations resulted in a legal action against the consortium in the united states District court of the Central District of California.

In the legal action, the plaintiffs sought damages for: [i] violation of the Racketeer Influenced And Corrupt Organization (RICO) Act, [ii] forced labor [iii] crimes against humanity; [iv] torture, [v] violence against women, [vi] arbitrary arrest and detention; [vii] cruel, inhuman and degrading treatment; [viii] wrongful death [ix] battery; [x] false imprisonment; [xi] assault; [xii] intentional infliction of emotion distress; [xiii] negligent infliction of emotion distress, [xiv] negligence per se; [xv] conversion; [xvi] negligent hiring [xvii] negligent supervision; [xviii] violation of California business and professions code at 17200; and [xix] injunctive and declaratory relief.

Further, there were campaigns in the United States against the joint venture. The campaigns resulted in ''selective purchasing laws" passed in the states of Massachusetts and Vermont, and in about 20 U.S cities. These law aimed at preventing states and city government from doing business with companies operating in Myanmar. The campaign and subsequent selective law caused a number of multinational corporations [MNCs] Pepsi-cola, Atlantic Richfield, Hewlett Packard, etc, to withdraw their investments from Myanmar.

CHAPTER FOUR

Some Impacts of Bretton Woods Institutions on the Extractive Sector

THE BRETTON WOODS INSTITUTIONS

The World Bank and the International Monetary Fund (IMF) are often known as the "Bretton Woods institutions". Both were established in 1944. The agenda for the Bretton Woods conference was to ensure that the world did not slip back into the chaos which had emerged after 1918. The scene for the conference was set in 1941, when President Theodore Roosevelt of the United States of America and Prime Minister Winston Churchill of Great Britain had agreed on the terms of the Atlantic Charter – to abolish colonialism, to develop a fairer world trade system, and to improve labor standards and security throughout the world. The Atlantic Charter was essentially the conditionality the US sought for its support of Britain's war effort. As the war drew to a close, the Bretton Woods Conference was to become the local content vehicle that ensured practical effect to the Atlantic Charter.

The Americans and British were determined to ensure that the post-war order would gain widespread acceptance, not only in their own countries, but also in the war-ravaged countries of Europe and East Asia, and in the countries that are about to be decolonized, into independence. The experience of the Great Depression, the demise of fascism in Europe and the military and economic triumphs of the USSR had made communism appear a very attractive alternative to capitalism and these were the "revolving gates" that gave Bretton Woods its ideological construction.

The World Bank, along with the International Monetary Fund (IMF), was established at Bretton Woods as part of the post-World War II international financial architecture. This system was meant to avoid future world wars by ensuring an open international trading system and global financial stability.

THE WORLD BANK

The World Bank's original purpose was to finance the reconstruction of war-torn Europe. Its early mission reflected the pressing concerns of the day which was captured by its original name known as the "International Bank for Reconstruction and Development (IBIRD)". The IBRD is now one of the Bank's two branches, the other being the International Development Association. In the post-war years, financing of physical infrastructure such as dams, highways, ports and railroads was the major priority, however, the World Bank has now become a major lending and technical agency concerned with development and for over the last thirty years, has increasingly become involved in debt-management in developing countries.

THE IMF

The IMF had a mission complementary to that of the World Bank. The severity of the Great Depression was aggravated by countries pursuing selfish competitive devaluations and fiscal contraction policies which can provide a reasonably assured but painful path out of a national recession, provided other countries aren't pursuing similar policies. The IMF's original mandate was to promote international monetary cooperation, to facilitate the expansion of international trade, and to promote exchange rate stability.

The IMF was charged with preventing the type of selfish fiscal management approach that could lead to another global depression. This the IMF ought to achieve by putting pressure on countries that were allowing their own economies to go into a slump, and therefore not contributing to global aggregate demand; helping those countries who could not afford to do so from their own fiscal resources and as well as their trading partners by providing governments with liquidity in the form of loans to countries facing an economic downturn and unable to stimulate aggregate demand from their own resources. But recently instead of the IMF to provide responsible economic stimulation for countries which could not afford to do so from their own fiscal resources, the IMF has come to be associated with fiscal stinginess, neglecting the very essence of its original charter, which was just the opposite.

Though the IMF key objectives remain sacrosanct, the methodology the IMF has engaged to achieve these has changed significantly, with increasing emphasis on imposing fiscal stringency on debtor nations.

While there is overlap between the World Bank and the IMF, there is a distinguishing difference which is that the World Bank's role has been economic development, while the IMF's role has been to prevent a recurrence of the circumstances which led to the Great Depression.

The International Monetary Fund (IMF), which started operation in 1946, promotes international monetary cooperation, exchange rate stability and the balanced growth of international trade. It provides member countries with policy advice, lending to support economic adjustment, and technical assistance. IMF technical assistance and training assist mainly low and lower middle-income countries to build human and institutional capacity for effective economic policymaking, focusing on such areas as monetary and fiscal policy, the exchange rate system, expenditure management, tax policy and administration, financial sector stability and statistics. Typically, this means that its poverty impacts are indirect.

The World Bank and the International Monetary Fund (IMF) together have more power to influence development in the developing countries than any other institutions in the world. The type of development influenced by these Bretton Woods institutions will continue for a long time to change the lives of those that benefit from their activities as well as dramatically alter local and global ecosystems.

IMF and World Bank practices have been criticised to have too often been characterized with taking over local development priorities, with their traditionally, short-term vision, insensibility to local state of affairs, and failure to consider the longer-term impacts of their activities on biodiversity.

The World Bank and IMF lending policies with its consequences have been severe on the developing countries. For instance, it has been insinuated that much of the Bank's US$22 billion annual lending supports projects and programs in environmentally sensitive areas, such as energy, agriculture and transport has been characterized by needless environmental destruction and missed opportunities for economically more efficient and environmentally friendlier alternatives.

In 1994, the World Bank internal review found that between 1986 and 1993, 15 per cent of World Bank lending was directed to projects that forcibly displaced 2 million people, the Bank closed or cancelled 22 of those projects, leaving 632,000 people to their fate. In the same 1994 the Bank approved 25 projects that forcibly displaced 458,984 people; a figure said to be the worst then than in any other year in the Bank's history. In contrast, however, in 1995 only 14 per cent of the Bank's outstanding loan portfolio was directed towards the Bank-defined sectors of "education" and "population, health and nutrition", while continuing to lend to new projects requiring forced resettlement. The Bank could point to only a few projects in which the affected public did not experience a diminished standard of living but all of the resettlement cases cited as successful were carried out under authoritarian governments.

The Bretton Woods institutions' policies and programs have often destroyed both the environment and the social framework in developing resource-rich countries it claims to be helping. The institutions have failed spectacularly to achieve their stated goals of poverty reduction and "sustainable development."

The IMF has the expertise but lacks the will to deal with social and environmental issues and operates within a framework that cannot accommodate the complexities of working with the widely differing economic situations in each country. The IMFs lack of comprehensive country-specific policies has impeded the long-term stabilization of developing countries' economies, calling into question the political feasibility of IMF structural adjustment programs, and, in general, the Fund's effectiveness in reaching its own economic goals. The IMF's continual disregard for people and the environment in borrowing countries has undermined the very foundations of sustainable development.

The Bretton Woods institutions recommended the Structural Adjustment Program (SAP) as the framework through which majorly developing countries can be economically and socially developed. Paradoxically, the SAP resulted in the weakening of institutional capacity and state sovereignty and also contributed to undermining the independence and legitimacy of the governments of mineral-rich developing countries.

The introduction of the SAP in the extractive sector and vigorously forced ingestion of SAP policy pills brought about the push to privatise and regulate the extractive sector, a shift away from employment protection and control of national natural resources to leaving the ownership, operations and management of the natural resources extraction to the private sector; it increased opening up of vast tracts of land, created a more favorable investment climate for foreign investors and subsequent vulnerability of the host and impacted communities.

The adoption of the SAP and its application in the extractive sector contributed to increased revenues generated by the extractive sector in the countries that have imbibed the SAP but such increase in revenue profile was more for the benefit of the investors and this was achieved through:

a. Exemption from duty payments;
b. Exemption from income tax payment;
c. Tax-free remittances for employees;
d. Scaled-down corporate income tax liability
e. Increased capital allowances;
f. Reduced royalty fees;
g. Scrapping of import duties; and

h.   Retention of foreign exchange earning in external account.

Further, the application of the SAP in the extractive sector created serious adverse social and cultural impacts in the host and impacted communities leading to:

a.   Persistent resistance by the communities and clashes with the extractive companies;
b.   Health and environmental problems;
c.   Destruction of sources of livelihoods;
d.   Displacements, relocations, resettlements, compensations and deaths;
e.   Loss of sources of water and firewood; and
f.   Reduction in agricultural activities due to encroachment and destruction of farm lands by extractive companies.

However, when the recommended SAP Program failed to yield desired outcomes, the Bretton Woods institutions were quick to blame the failure of the SAP on internal factors such as corruption, lack of transparency and 'weak governance' in the developing countries, instead of blaming the not well thought through SAP policy framework. The Bretton Woods emphasis on corruption and lack of transparency in the developing host governments nonetheless has emasculated the fact that such situations are often facilitated and even perpetuated by the extractive companies and their home states through Bilateral Investment Agreements (BITs) to arm-twist local decision makers and political leaders.

But is eradicating corruption the real healing drugs for the surreptitious unsustainable activities of the extractive industries with its manifest linkage effects to bloody wars, human rights abuses and ecosystem destabilization or the Bretton Woods institutions actually misdiagnosed again. Corruption is an issue of unquestionable concern here because according to the European Commissioner for Home Affairs, Ms Cecilia Malmström, who revealed in Brussels in early February 2014 that "corruption in the 28 countries of European Union (EU) is costing European taxpayers about £100 billion a year, the equivalent of the Union's Annual budget".[15]

Yet with this magnitude of corruption among members of the EU, extractive companies and governments operating in the Euro zone respect the rights of their citizens and do not intentionally destabilize the ecosystem unless by accident and when the accident does occur, there is always an immediate response to salvage the situation and curtail the level of human and environmental damages. But this is hardly so in developing world where the

---

[15] Olatunji Dare, Corruption: The EU to the rescue (The Nigerian Nations Newspaper Tuesday February 11, 2014) page 64.

same extractive companies operate. Therefore is it actually the case of corruption or confusion or even outright deceit and bias to the developing countries extractive jurisdictions?

We now visit few countries where these Bretton Woods institutions have had the opportunity to work with government and see the immediate outcome of their intervention. We shall now localize our general observations to few selected countries that adopted the SAP in their extractive sector, starting with DRC.

THE DRC EXPERIENCE WITH THE BRETTON WOODS INSTITUTIONS

The principal roles of the Bretton Woods institutions in the DRC are salient in the mining sector, since the years of the imposition of structural adjustment programs as standardized development programs based on new regulatory approaches in an attempt to deal with the country's structural poverty. The structural Adjustment Program in the DRC like in every country which they are directed for implementation has had compounding negative impacts albeit with some successes.

From 2001, the World Bank and IMF have influenced development programs to be restructured in the DRC through a synchronized framework of actions ensuring 'world partnership for development', captured in the MDGs.

The desire to support and collaborate with the Congolese government resulted in the mobilization of partners and various funding sources intended to create convergence between the various development initiatives. This mobilization was provisional on the following three issues:

- Pursuing further state reformation strategy;
- Adopting international development agendas and standards;
- Democratizing and stabilizing the political environment.

Through the massive support of foreign funding agencies, a new Constitution was adopted in February 2006, based on the constitutional referendum held on 18–19 December, 2005; and in October 2006, the presidential election took place which saw Joseph Kabila emerge as president. As of March 2007, the World Bank committed itself to invest US$1.4 billion in the DRC over a two year period (2008–2010), in the form of grants and loans. The World Bank and the United Nations together developed in September 2006 a strategic document called *Country Assistance Framework* (CAF) as the 'common strategic framework' to support the DRC and this document was to become the reference framework for all development programs supported by the European Commission, the United States, China, the United Kingdom, Belgium, Germany and France. The declared objective was to enable the DRC to achieve the MDGs as soon as possible, most probably within five years. The *Framework* was structured around these five objectives:

- Effective management;
- Pro-poor growth;
- The provision of basic social services like education, water and health;
- Providing a social safety net;
- HIV/AIDS treatment and community development.

The initiative led to extensive restructuring of state institutions and the treatment of development issues in a more technical manner aimed at reinforcing the institutional, legal and management control systems to ensure 'good governance' or 'effective management' practices to strengthen the rule of law, put an end to the economic crisis and ensure repayment of outstanding debt, as the beacon of hope for the new development agenda.

These reforms transferred state undertakings to the private investors, thereby reducing to the minimum the production and distribution functions of the state, through a dual logic of decentralization and privatization. Such a transfer was made possible by the enabling environment created through the Bretton Woods institutions-led reforms that opened up the sector to foreign private investments, liberalization, leading to the progressive dismantling of state companies, reducing the financial burdens borne by companies and reinforcing administrative control mechanisms, with the espoused purpose of fighting against corruption and promoting transparency.

With the reform agenda, the private investors now have more powers, becoming the holders of mining rights, investors, operators, buyers and sellers, and taxpayers. The state ceased its functions as a producer, retaining above all only the function of facilitating investment and providing security to the sector through its national security personnel and apparatus, with its revenues generated through taxes, fees and royalties paid from the exploitation of natural resources by mostly privately owned foreign capital investments. The state also has to guarantee a fair redistribution of mining revenues among the different sectors of the population, through decentralization, within a legal framework, to provide for basic services such as education, health, water and sanitation, housing and so on and so forth.

On their part, the extractive industries are to respect the legal frameworks, to guarantee that they will continue to generate revenue for the state and, on a voluntary basis, to commit to resolving certain negative externalities, such as pockets of poverty around mining sites or environmental pollution, through its corporate social responsibility framework. In all major ramifications the WB and IMF backed SAP favor the promotion of private property rights.

The WB and IMF led reforms in the DRC like in other developing countries overburdened under the heavy weight of SAP have not met the needs of the country; have not led to economic progress that is respectful of the ways of life and interests of the host communities; and have not led to sustainable development.

THE GHANA EXPERIENCE WITH THE BRETTON WOODS INSTITUTIONS

Afraid of being declared insolvent by the IMF due to severe depression in 1983, the Ghanaian government entered the IMF Structural Adjustment Program in that year. At the beginning of the Program, the World Bank specifically informed the Ghanaian government that "the forestry sector offers the greatest immediate potential for growth and foreign exchange earnings" (World Bank, 1984 as cited in Owusu, 1998).

The forestry sector received one of the largest sector adjustment loans in order to replace worn forest equipment in preparation for its coming role as the primary source of foreign exchange. As one of the conditionality to access the IMF facility, Ghana removed some of its export restrictions scheme. Furthermore, to ensure high export performance, the Timber Export Development Board (for export promotion) and the Forest Products Inspection Bureau (to monitor production) were established to replace the Ghana Timber Marketing Board. Finally, the local currency was devalued to enable the government to continue to pursue higher export levels in an attempt to maintain the stability of its hard currency revenues.

By 1993 the volume of lumber exported from Ghana had increased by 500% while the volume of logs exported increased by 806%. According to the Timber Exports Development Board, and the Forest Products Inspection Bureau, between 1983 and 1991, the total foreign exchange generated by the export of wood and wood products increased from US\$15.77m to US\$114.2m. The Ghanaian government was then compelled by the IMF to use the revenue generated from timber and log export to service the interest on its foreign outstanding external public debt. Ghanaian government used "desperate deforestation" in order to satisfy international capital and return to normal relations with Ghanaian creditors (Owusu, 1998, p. 428).

The International Monetary Fund (IMF) officials were acutely aware that natural resource degradation that threatens growth cannot be ignored. According to the First Deputy Managing Director of the IMF, Stanley

Fischer, poor environmental conditions can have an adverse impact on economic growth and macroeconomic balances[16].

Also, IMF officials contend that their Programs have beneficial effects on the environment because the macroeconomic stability their Programs promote is vital for environmental preservation (Fischer, 1996).
Critics of IMF Programs have claimed that the Fund policies hurt the environment by encouraging budget cuts to environmental Programs, promoting primary product export-oriented development, and inducing economic contractions that lead to extensive migration to marginal lands. But the question that continues to agitate fair-minded public is whether IMF policies and programs actually have negative consequences on the environment which is the linchpin towards the achievement of sustainable development.

Could there be any relationship therefore between IMF Program and environmental degradation? This question was answered with a study by James Raymond Vreeland, Robynn Kimberly, Sturm Spencer and William Durbin of the Yale University who used a data set of 2,258 observations from 112 countries from 1970 to 1990 to conduct a study that actually confirmed the fear of IMF critics that IMF Programs cause deforestation increases especially when governments participate in the Programs.

Despite the degree of disagreement over the negative consequences or otherwise of the impacts of IMF Programs on the environment, there is quite some evidence linking IMF Programs to environmental degradation through raw materials exports, mineral depletion, and deforestation, among others.

If environmental degradation is linked to IMF through its Programs, could it not be wise then to conclude that IMF Programs cannot lead to sustainable development due to the consequences of deforestation on the environment, global warming, devastation of the biodiversity network, flooding and erosion; unsustainable livelihoods sources, air, land and water pollution.

According to Wilfrido Cruz, an associate at the World Resources Institute, "The deterioration of a nation's natural resource endowment is at least as serious an obstacle to sustainable development as the deterioration of its international credit standing"[17].

---

[16] Fischer, Stanley, 1996. "What is Reasonable to Expect of the IMF on the Environment?" in Gandhi, Ved P., ed. *Macroeconomics and the Environment* Washington D.C.: International Monetary Fund, pg. 248.

[17] Wilfrido and Robert Repetto, 1992 "*The Environmental Effects of Stabilization and Structural Adjustment Programmes: the Philippines case*". Washington, DC: World Resources Institute pg. 67.

The Assistant Director of the Fiscal Affairs Department at the Fund writes, "Ignoring environmental degradation means ignoring its impact on human capital, natural capital, and output, all of which have a bearing on the sustainability of macroeconomic stability and economic growth".

Entering into IMF Structural Adjustment Program, means agreeing to some conditionality imposed by the Fund which include removing impediments to export growth, lowering tariffs, currency devaluation, removing subsidies and cutting government expenditures.

The IMF policy conditions of removing impediments to export have detrimental effects on the environment through excess logging leading to increased rate of deforestation, if that country's economy is mostly centered on timber export. Removal of subsidies leads to increased spending on critical products and services needed to sustain livelihoods and wellbeing. In this case people may resort to activities leading to deforestation as source of revenue to augment their income and maintain humanistic life style. Further, devaluation of the national currency may also increase incentives to cut down forests, as devaluing the currency effectively lowers the price of forest products on world markets, thereby increasing demand. Also under IMF agreements governments are required to raise foreign capital to repay the IMF loans and this has led to promoting exports, including forest products.

Furthermore, cutting government expenditures may adversely affect the environmental protection and enforcement Programs which often lose funding when government budget deficits are reduced. Officials from the IMF and the World Bank recognize this possibility. The World Bank Environment Director, Andrew Steer, notes, "this category of public expenditure [environmental Programs] may be cut as much or more than other categories of expenditure" (Steer, 1996, p. 67). In some countries, environment departments are important mechanisms for helping companies manage sustainable forests, curtailing illegal timber harvesting and educating individuals about deforestation. When these Programs are cut, deforestation is apt to increase.

IMF policy conditions also leads to deforestation through unemployment and income decline which forces population shifts from urban centers to subsistence living in rural areas, promoting conditions for the poverty-stricken population to overexploit the environments.

THE MALI EXPERIENCE WITH THE BRETTON WOODS INSTITUTIONS

The engagement of Mali with the Bretton Woods institutions is very important to mention in this book. Mali's economy for a time was based principally on cotton, rice, fruits and vegetables, food gathering and cattle and mining of gold. The country was for a long time the continent's largest producer and exporter of cotton. However, by 1997, gold displaced cotton production in Mali, and the country ranked fourth among African countries producing gold. In spite of huge revenues derived by Mali from gold, the country for several decades has remained one of the least-developed countries according to international financial institution (IFI) criteria.

To reverse the opaque poverty situation in the country, Mali imbibed in the 1980s liberalization policy recommended by the Bretton Woods Institutions for its mining sector, which was aimed at re-establishing macroeconomic equilibrium, control debt, ensure market reform and overhaul the mining framework for private enterprise investments. This move was to ensure state's withdrawal from direct mining activities and to focus on regulation of the mining sector, and the promotion of private initiative as the vehicle to stimulate socio-economic development of the country. This liberalization policy meant development of the legal and regulatory system that will provide a framework to manage the economic activity of the country in such a way that will catch the attention of foreign investors and promote the private sector; to position the mining sector as the indispensable focal point for economic growth, livelihoods improvement and poverty reduction.

The funding agencies encouraged Mali government to undertake this reform activity which not only provided more incentives to but benefitted the foreign mining companies more than the host government, through less government participation in the capital of mining companies, lower tax regime in that mining companies were excused from paying taxes during the first five years of production until 1999 when the tax-free regime was removed, permanent exoneration from payment of customs duties and taxes, exoneration for three years on Value Added Tax, and accelerated depreciation of their factors of production, among other incentives. This made the economic and financial regulatory framework inordinately eye-catching to the foreign mining companies.

With liberalization and privatization of the mining sector, mining projects in developing countries have increased pressure on wood resources and also cause global warming in the mining regions as a result of deforestation that depletes carbon sinks. Also, the communities living around the mining zones lack basic infrastructures and social services. Though some mining companies build public health clinics for mine workers and provide equipment for community health center as part of their corporate social responsibility, the local host communities' lack of funds limits their access to healthcare services provided by the mining companies. Furthermore, the facilities created for mine employees are not accessible to the rest of the population.

In general, the application of the regulatory framework and the weak participation by national actors, particularly the state, has contributed to creating tensions between the companies and the communities in which they operate. The communities receive little information about the procedures and results of environmental management.

Prior to the liberalization of Mali's mining sector, Mali adopted the World Bank and International Monetary Fund recommended Structural Adjustment Programs in the 1980s that eventually like in all other jurisdictions that adopted the same SAP framework, led to weakening of the technical and financial capabilities of the country to manage its mining sector thereby preventing the mining sector from living up to the bidding of becoming the country's focal point for sustainable socio-economic development.

The SAP had two major negative effects on the Malian mining sector. First, monitoring of the mining sector was carried out by the staff of the Direction *nationale de la géologie et des mines* (DNGM), the same agency that also was in charge of geological research. However, the SAP led to substantial personnel cutbacks in the public service and reduced the possibility of hiring more civil servants, including the staff of DNGM who are well qualified and experienced to ensure effective operation of the mining sector, thereby depriving the ministry and particularly the DNGM of the services of those staff who were most competent to oversee the mining sector. These sacked mining staff of DNGM, once the mining companies started their operation, became ready source of recruitment of capable and experienced staff for the foreign mining companies.

Another requirement of the SAP in Mali was that the Malian government created a single centralized fund for the state budget, which meant stifling the various funds used to encourage the industrial sector growth and depriving the mining ministry of access to essential funds which had in the past served to finance its own activities. Under these conditions, this ministry, as well as the operational structures charged with monitoring and controlling the mining industry, found themselves seriously limited in both the human and financial resources essential to carry out the role assigned to them by the country's legislation.

The impact of this is that the country had difficulties in assessing the financial data provided by the mining companies regarding their levels of investments, operating costs and productions and the capacity to verify that the revenues it receives from the mining companies were appropriate as it had to depend on the figures produced and supplied by the mining companies.

The inability of Malian government like other governments that hitherto imbibed the SAP to ensure adequate monitoring of mining activities in the country became a major inhibition of the country's mining sector as contributor to poverty reduction and the protection of the environment.

THE GUINEA EXPERIENCE WITH THE BRETTON WOODS INSTITUTIONS

Guinea is blessed with abundant high-grade solid mineral called bauxite and the country is considered the world's most important source of this mineral. The World Trade Organization (WTO) in the year 2005 estimated that Guinea bauxite reserves was about 20 billion metric tons and are of exceptional grade in both quantity and quality. Also it was believed then that the country accounts for about 40 per cent of world trade in this mineral, and the supply of equal amount to the United States[18].

Despite the country's potential for wealth generation from its rich mineral sector, the impact of the mining sector on the Guinean economy was nothing to write home about because after independence, resource extraction became far less favorable to the country than it ought to be as revenues from the mining sector started to decline after hitting a record of over 93 per cent of the country's export earnings.

The World Bank and the IMF carried out detailed studies on the Guinean mining sector and its impact on the economy which resulted in a preliminary discussion with the President Sékou Touré administration on a possible structural adjustment loan; however, little progress was made before the death of President Sékou Touré in early 1984.

The death of President Sékou Touré was followed by a bloodless military coup that brought General Lansana Conté, the coup leader, into power as the President of Guinea. President Conté who was very enthusiastic to put into practice economic reform, contacted the IMF and World Bank just seven days after his takeover of Guinean government. Under President Conté, Guinea government adopted the World Bank recommended SAP.

In response to the declining revenue from the mining sector in the 1980s and 1990s, Guinea Mining Code was revised in June 1995, with recommendation and backing of the World Bank. Like in Mali the new mining code harmonization Guinea mining sector according to the World Bank, as a measure to provide new incentives in order to attract foreign investment. The Guinea mining regulatory framework was also reformed especially in areas of conditions of employment and the repatriation of profits and this was believed again to allow for significant boost in the amount of bauxite and alumina produced and a rapid expansion in the extraction of gold and diamonds during the decade to follow.

---

[18] World Trade Organization (WTO) (2005) *Trade Policy Review: Republic of Guinea – report by the Secretariat (revision)*, Report No. WT/TPR/S/153/Rev.1, Geneva: Trade Policy Review Body, December 14.

The disappointing results of the Bretton Woods institutions' recommended Structural Adjustment Programs and liberalization of the mining sector explicitly weakened the institutional capacities of Guinea to effectively manage its mining sector in the interest of the general public.

Though there were some problems here and there with the implementation of the structural adjustment reforms, however, there was a very significant shift towards a harmonization economy as of mid-1986. In 1988 and again with the recommendation of the World Bank a second structural adjustment Program (SAP) was designed and adopted to deepen and consolidate the reforms of the first SAP. This second SAP was expected by the World Bank to address the 'domestic policy inadequacies' of the previous regime, and greatly improve the country's economic performance. These projections, however, failed headlong before it could hit the runway.

Towards the end of 1985, under the leadership of the IMF and the other international lending agencies, the government of President Lansana Conté, the Comité militaire de redressement national (CMRN), adopted a two-year Program intérimaire de redressement national (PIRN) which had been conceived to bring economic redress and re-establish balanced public finances. The Program had five major aspects:

- Monetary reform;
- Liberalization of foreign trade and domestic prices;
- Privatization or liquidation of the majority of state owned ventures;
- Administrative reform;
- Legislative reform (particularly affecting business, commerce and new investment).

The PIRN was replaced in 1985 by the Program de réformes économiques et financières (PREF), designed to transform Guinea from a centrally planned to a market economy. This Program, which closely mirrored the World Bank and IMF recommendations, allowed for the rescheduling of Guinea's huge unpaid external debt in April 1986 and the introduction of a first structural adjustment loan.

Agreement was also reached with the IMF, and a three-year Structural Adjustment Facility (SAF) credit of Special Drawing Rights (SDR) of 36.8 million was endorsed in 1987. The second phase of the structural adjustment Program (SAP II), launched in mid-1988, continued with further reforms of the management of public finance and accelerated restructuring of the civil service and public sector. The SAF expired in mid-1990, but Guinea was unable to negotiate a new facility with the IMF until November 1991. The reason for this delay was Guinea's failure to meet a number of criteria written into the original SAF agreement, notably delays in the privatization of the state fuel Distribution Company and failure to reduce the size of the civil service.

In November 1991, a Three-Year Program (1991–4) was approved by the World Bank and an Enhanced SAF (ESAF) agreed upon by the IMF conditions attached to the ESAF required Guinea to:

- Further reduce the size of the civil service;
- Continue the privatization Program;
- Reduce public utility subsidies;
- Respect promises of public spending restraint.

As with many previous funding arrangements, the ESAF was blocked for several months because of unsatisfactory budgetary performance which in turn delayed funding from the World Bank and the Paris Club, which the IMF must approve as a prerequisite to reopening credit lines and consideration of debt rescheduling.

To add to the country's problems, Guinea's budget for the year 1993 which was supervised by the IMF was extremely difficult to achieve because of the unrealistic revenue projections on which it was based, notably with respect to fiscal receipts from the mining sector.

The anticipated outcome was not long in materializing. Government revenues in 1995 proved to be very much below expected levels. Consequently, after having approved a third and fourth instalment of the ESAF of SDR57.9 million (US$86 million), which was to have been made available in November 1996, the IMF decided to suspend its funding until after the meetings planned with the country in the autumn of 1996.

Among the various measures proposed to remedy the weakness of public finances, multilateral financial institutions insisted on the improvement of the management of customs services, the introduction of a value added tax of 18 per cent and the liberalization of regulations concerning the mining sector in order to attract new investors.

Above all, however, the decrease in revenues from bauxite and alumina placed extremely severe constraints on the country's financial operations. But instead of searching for additional or alternative measures to mitigate the constraints, the response of multilateral financial institutions, and notably the World Bank, was to introduce a 'rescue plan' which took the form of the new 1995 Mining Code. This plan failed, however, to take into consideration the reasons for the stunning drop in mining receipts over the preceding ten years. In the absence of such consideration, the strategy proposed appeared to focus more on attracting increasing flows of foreign investment on increasingly harmonization terms.

However, ten years after the above reforms which were undertaken to liberalize the mining sector to attract foreign investment and boost the Guinean economy, the mining sector failed to produce the desired positive impact on the national economy. In fact the entire experiment was disappointing from the standpoint of the well-being of the Guinean people.

This disappointment and its enormous social and economic consequences led to the popular uprising of January 2007 that paralyzed the country and the government in a widespread social mobilization against the government, under the leadership of two unions - l'Union des travailleurs de Guinée and the Confédération nationale des travailleurs de Guinée, due to soaring degree of poverty and frustration experienced by Guinea citizens. The state of affair this period in Guinea threatened the authority of the President Lansana Conté led government and paralyzed the function of the government. Confronted by this popular unrest, President Lansana Conté countered with sadistic despotism enforced through the police and the presidential guard. At the end of the imbroglio, two major achievements were recorded in favor of the protesting unions; the first being the nomination of Lansana Kouyaté, favored by the unions as Prime Minister and the second was the announcement in April 2007 that mining contracts previously signed between the government and foreign mining companies would be reviewed.

THE EL SALVADOR'S EXPERIENCE WITH THE WORLD BANK TRIBUNAL

Salvadoreans opposing a planned exploration by Canadian mining firm Pacific Rim protest in front of the Canadian consulate in San Salvador, on May 21, 2007. The sign reads "Yes to Life." Jose Cabezas/AFP/Getty Images

In April 2014, more than 300 international and national civil society organizations signed and sent a protest letter to the World Bank president, Jim Yong Kim, during the bank's biannual meeting in Washington, criticizing the bank's involvement in the case of *Pac Rim Cayman LLC v. El Salvador*. Pacific (Pac) Rim Cayman LLC which is a transnational mining corporation acquired by Canadian-Australian Oceana Gold sued the government of El Salvador for compensation for lost of investment and future profits, alleging failure of the El Salvadorian government to approve an extraction permit after the company has purportedly invested millions of dollars in the exploration of the El Dorado mine in the north-eastern province of Cabañas.

The controversy has ignited a debate over whether disputes between countries and corporate investors should be adjudicated in national courts or international tribunals.

In this suit that was first filed in 2009, El Salvador claimed the company failed to follow proper protocols for issuance of a license. According to the El Salvadorian government, Pacific Rim did not possess title to much of the land considered for the mining project, also failed to secure the appropriate environmental authorizations and never submitted the final feasibility study for the project.

Instead of filing the suit in the national courts, Pacific Rim lodged its complaint at the World Bank's investor protection tribunal, asking for more than US$300 million, almost 2 per cent of El Salvador's gross domestic product (GDP), in compensation for spending on exploration and for loss of future profits.

The Labor Union, the grassroots and human rights organizations argue that it is inappropriate for the World Bank, whose mission is alleviating poverty, to preside over disputes that threaten the self-determination of countries. All these more than 300 international and national civil society organizations were protesting against Pacific Rim's dependence on the World Bank's arbitration mechanism to subvert local authority over tangible issues germane to the well-being of poor communities, which should take precedence over profits for transnational companies.

This suit was in June 2012 dismissed by *the International Center for the Settlement of Investment Disputes* (ICSID), an arbitration mechanism under the Dominican Republic–Central America Free Trade Agreement, for lack of jurisdiction, but allowed the suit to proceed under El Salvador's investment law, which at the time provided for international resolution of disputes, but has since been amended to ensure that disputes are adjudicated in national courts instead of international tribunals, unless dictated by bilateral trade agreements.

The investor arbitration provisions in free trade agreements and bilateral investment treaties (BITs)[19] pit the interests of transnational capital against national sovereignty, economic self-determination and sustainable development. BITs have become the best mechanisms to shield the transnational corporations in the extractive sector in poor developing countries of the world. These extractive industries are often supported by Western neoliberal economic policies and international financial institutions, such as the World Bank and International Monetary Fund, that finance and direct development initiatives that demand privatization and free trade.

The Latin American countries have also had their disproportionate share of the tactic of these Bretton Woods institutions. As of March 2013, Latin American and Caribbean countries comprised only 14 per cent of the 158

---

[19] See Chapter ------ on full discussion on BITs

ICSID member states, but were responsible for 46 per cent of the ICSID docket and more than 50 per cent of the pending cases involving the extractive industries[20].

The World Bank believes it offers a most consensual neutral cost-effective dispute resolution mechanism through mandated arbitration framework built into trade and investment agreements. However this arbitration mechanism has compelled impoverished countries to submit to its jurisdiction. It is argued here that the bank's removal of community and national control over local projects threatens to further impoverish marginalized communities mostly impacted by extractive projects.

The people and government of El Salvador are afraid the proposed mine will pollute and threaten their already contaminated water supplies and degrade the environment. They are against mining projects that will ravage aesthetic beauty of their landscape, degrade their environment and compromise their health.

The assertion is supported by the fact that former Presidents Tony Saca and President Mauricio Funes refused to issue a mining permit to Pacific Rim. Further, the President-elect Salvador Sánchez Cerén vowed that no permit should be issued to Pacific Rim.

El Salvador's struggle to maintain control over its development is being keenly watched. The world and especially the developing countries where extractive transnational corporations operate have watched this development with eagle's eye as this can trigger some measures to reduce the stranglehold of the multinational extractive corporations on developing countries, some of which have now realized the need to fend off the unsustainable exploitation of their natural resources and protect their national environmental health.

The outcome of this dispute will set an important precedent about the reach, legitimacy and costs of investor protection provisions in existing and future trade and investment agreements.

Stripping nations of their right to economic and environmental self-determination by privileging investors' rights through international arbitration mechanisms could be adjudged as an infringement by foreign private investors on the sovereign powers of States. This encumbers the rights of host communities and nations from benefiting from the development of natural resources.

---

[20] Lauren Carasik, *"Organization's investor protection panel disempowers marginalized communities"* (Aljazeera News April 22, 2014)

The decision of the World Bank to allow Pacific Rim to avoid local dispute resolution mechanisms and challenge the sovereign power of a nation to take decisions on its socio-economic development in international fora may a threat to the spirit of sustainable extractive sector management.

CHAPTER FIVE

Some Attempts towards Sustainable Extractive Sector Management

In this chapter, the book concentrates on a few out of many various efforts made by various key stakeholders in the extractive sector to reduce the extant acrimonious relationships that exist among development companies, governments and communities. Though many efforts have been made along this line, however, only few will be presented here and they include:

1. World Bank-Netherlands Partnership Program (BNPP) Trust Fund for Women's Access to Resources and Voice in Papua New Guinea;
2. The World Bank Energy, Environment, and Population (EAP) Tripartite Dialogues Initiative in the Sub-Andean Basin;
3. The World Bank Role in the Extractive Industries Transparency Initiative;
4. How the United Nations is Contributing to Sustainable Extractive Sector;
5. Chevron's Activities towards Sustainable Development in Niger Delta.

The presentation starts with the World Bank-Netherlands Partnership Program (BNPP) Trust Fund for Women's Access to Resources and Voice in Papua New Guinea

1. WORLD BANK-NETHERLANDS PARTNERSHIP PROGRAM (BNPP) TRUST FUND FOR WOMEN'S ACCESS TO RESOURCES AND VOICE IN PAPUA NEW GUINEA

"When the women begin to take an active part in the struggle, no power on
earth can stop us from achieving freedom in our life time"

Chief Luthuli of South Africa

As indicated above, in the extractive host communities, women are particularly harmfully affected by the negative consequences associated with the mismanagement of extractive industries. The women are discriminated against and denied the opportunity to participate in decisions on the processes and strategies aimed at transforming the extractive industries into visible social and economic benefits in affected communities. This observable fact is unfortunately ever present in almost every extractive jurisdiction in developing countries the world over. To reverse this trend therefore, women voices need to be heard and their capacity developed to enable them effectively engage in decisions regarding the management of the extractive sector and transparent utilization of revenues generated from the sector.

The World Bank has duly recognized this need and is actively initiating or encouraging positive strident activities to ensure women actively participate in decisions on the management of the extractive sector. One of such efforts among many is the World Bank-Netherlands Partnership Program (BNPP) Trust Fund to develop knowledge and learning tools in Papua New Guinea (PNG), which mobilized women in the mining sector and developed their coalition-building strategies and approaches; their interactions with mining sector institutions; and their overall strategies for accessing resources, knowledge, finance, and capacity-building opportunities.

Drawing from practical examples and lessons, the strategies builds on the Women in Mining (WIM) initiative and the analytical work on the special provisions of the Community Mine Continuation Agreement (CMCA) supporting women in the CMCA regions of Western Province of PNG. The WIM initiative, CMCA Women and Children's Action Plans, and the CMCA negotiations provide positive examples of World Bank initiation or encouragement of women-driven engagement with government and private sector actors aimed at improving the living conditions of their communities.

With the hindsight of the challenges facing women participation in decisions regarding development in their communities and the lessons from their participation in the negotiation process, the project ensured that women in PNG participated in negotiating mining operations' benefit streams for local communities. The project was adjudged relevant for PNG and for other countries in their attempts to make policy decisions about transforming natural resources wealth into inclusive and sustainable development for local communities.

The participation of PNG women in the negotiation was first implemented in 2006/07 where women in the nine CMCA regions of the Western Province of PNG were given the opportunity to participate in the negotiations for mine benefit streams for communities affected by the operations of Ok Tedi Mining Limited (OTML). The women were represented at the negotiations table by only one woman. At the 2006/07 negotiation, through the dogged effort of the women representative, the women were able to secure a 10 per cent of the funds set-aside from the mine operations to be dedicated to women and children's development programs. The 10 per cent trust fund translated into PNG Kina (PGK) 101 million (about US$38 million).

Apart from securing the 10 per cent women and children development trust funds through women direct participation in the negotiation, this involvement reinforced women's access to rights of representation at the highest levels of decision-making on mine benefits for local communities.

The Memorandum of Agreement (MoA) that resulted from this negotiation round specifically provided for the recognition of women representatives on Village Planning Committees (VPC), on the CMCA Association, and on the Board of the Ok Tedi Development Foundation (OTDF). The 2006/07 MoA also stipulated that the MoA would be reviewed every five years.

After first five years, the agreement came up for review in 2012 and during this 2012 negotiation, up to 30 women leaders participated in the negotiations process, which was yet another pioneering milestone with not only the increase in the number of women participants, but also the women's ability to increase their set-asides for women and children development programs from 10 per cent achieved in 2006/2007 to 18.24 per cent in 2012.

This success in the 2012 negotiation round was achieved because, with the 10 per cent set-aside in 2006/07, the women were able to mobilize, plan, and build their capacity; worked hard enough to demonstrate some benefits from the training that they had received through the use of the allocated funds. This development emphatically gave the women the propulsion to be more explicit in their argument in 2012; benchmarking their presentation of the fact that they were not acting only on their own interest but on behalf of their communities.

Though the women were excited with their successful outing in the 2006/07 CMCA negotiation round; however they were still confused about how to access their 10 per cent development fund or what it all means. And life has remained the same as it was before they secured the deal as the impacts of the 10 per cent funding were yet to be felt by women in general, in the mine-impacted communities. It is therefore important for the women in the villages to be organized and empowered through information dissemination and increased capacity to upgrade their decision-making powers.

This will facilitate the reversal of the challenges by strengthening the women capacity to fill their deficits and gaps in knowledge, especially limited understanding about important issues such as the content of the Memorandum of Agreement; the sources of funding; the roles and responsibilities of the different entities charged with implementation (CMCA Trusts, PNGSDP, OTDF); and the rules and procedures for accessing the funds which have restricted their ability to drawdown on the 10 per cent set-aside funds. Two additional challenges stand out:

If the capacities of the women are strengthened they would be able to achieve the needs they identified during the 2006/07 negotiations, such as training in cooking and sewing, computer skills such as how to use PowerPoint presentations, and the purchase of the multi-purpose MV Fly Hope (a multipurpose boat for river transport). Also, they would have been able to understand that it was not just enough to have access to their identified needs, they must have access to the most critical enablers to effective service delivery (such as roads, jetties, and bridges; health centers and outposts; classrooms, libraries, and accommodations for teachers, nurses, and doctors) and the elements of sustainable livelihoods (such as agriculture and food production) that the women enumerated in their action plans. These benefits are yet to be fully realized.

Further, building the capacity of the women will strengthen representation on the OTDF Board which has fallen short of expectations as there was only one woman on the OTDF Board. Aside from the fact that this is an unfulfilled legal requirement, the real issue is about improving OTDF's performance and promoting equal opportunity for CMCA women. The business case for CMCA women on the OTDF Board goes beyond women's quotas to encompass promoting women's voice and ensuring that different perspectives are brought to the negotiating table and debated or considered in a holistic way. The women leaders represent the voices of the marginalized and underprivileged people back in the villages.

The experimentation carried in PNG by the BNPP provided important opportunity for the CMCA women leaders to be factored into decision-making process in natural resources management and built their capacity to recognize their needs for:

- *Capacity development.* Institutional capacity and human resource development as the primary means of ensuring that women are able to manage their Associations, take control of the Women and Children's Funds, and manage them separately from the current Trust arrangement;
- *Stakeholder collaboration.* Closer collaboration and partnership between all stakeholders (including OTDF and PNGSDP) in order to complement each other's efforts in delivering credible projects;
- *Empowering Village Planning Committees with project management skills.* Empowering Village Planning Committees to manage small village projects that ensured project ownership. Women leaders want

OTML to start working first with those Village Planning Committees that have a proven track record of decision-making for village projects and are accountable and transparent in participation processes;

- *Ownership and sustainability.* The negotiators wanted women leadership to be consulted to ensure ownership and sustainability.

From the activities of the BNPP in PNG it is hereby rightly concluded that this engagement made significant strides in securing women's access to voice, representation, and rights of participation in the management of the extractive sector. This success is attributable, in large part, to:

- *Higher levels of education attainment which enhances women's leadership role.* Through the interviews conducted, it became clear that those who spoke out were women leaders who have attained 10th Grade of education and above and had some technical training.
- *Level of preparedness and relationship building played effective role during negotiations.* Most of the women leaders had continuously attended all regional MLE meetings over three years. They were familiar with the issues in addition to knowing the OTML Community Relations Officers through the prolonged periods of engagement.
- *Community interests (regional) often take precedence over local (village-level) positions.* In general, women tend to identify specific needs more at their locality than at the regional level. However, during the negotiations, the community interests at the regional level took precedence over village-level interests.

The challenges identified with this project are how to translate the regional agreements into tangible benefits for everyone in the community. These include:

- *Changes in attitudes and mind-sets are happening, but the pace is not fast enough.* Although there is some evidence that mind-sets and attitudes are changing, such changes are not happening fast enough in the communities that are impacted most by the OTML operations. Social and economic development is lagging and indicators of women's overall empowerment on such issues as maternal mortality continue to be low, especially in the Middle and South Fly regions.
- *Implementation of the 2006/07 CMCA remains a challenge.* As the CMCA-negotiated funds have grown bigger, emphasis should now be placed on how and on what to wisely spend the negotiated funds. The CMCA Women's and Children's Action Plans 2009–2019 offer an opportunity to dictate and drive the Budget for Priority Expenditures identified by the CMCA women.

2. THE WORLD BANK ENERGY, ENVIRONMENT, AND POPULATION (EAP) TRIPARTITE DIALOGUES INITIATIVE IN THE SUB-ANDEAN BASIN

The activities of the extractive industries resulted in extremely challenging economic, social and environmental problems in the Andean countries (Bolivia, Colombia, Ecuador, Peru, and Venezuela) of the Amazon Basin, where oil and gas development take place. The governments of these Andean countries have the responsibilities to enforce environmental, social, and human rights legislation, and international treaties such as the International Labor Organization (ILO) Convention 169, which provides for the protection of the rights of indigenous peoples, as well as their own national regulations concerning free, prior, and informed consultation with indigenous populations.

However, there was a huge challenge for these governments to meet the obligations of effectively managing their extractive sector in line with the extant regulatory frameworks. Of course what followed was rancorous relationship among the three critical stakeholders: the Andean nations, the oil and gas industry, and indigenous organizations.

In response to these challenges facing the governments and the rancorous relationships that exist among these critical stakeholders, the World Bank intervened by creating the *Energia, Ambiente, y Poblacion* (Energy, Environment, and Population (EAP) Program as a trust- building initiative that will help them better understand and address each other's needs. The Program which was initiated and developed to address the challenges of the oil and gas industry in the culturally and ecologically sensitive Amazon rain forest Basin, which is one of the largest River Basins in the world, provided an innovative forum for tripartite dialogue and consensus building over the development of the hydrocarbon sector in these countries.

The EAP initiative was inspired by the "Harvard Dialogues on Oil in Fragile Environments" developed during the 1990s, which aimed to promote better understanding among the oil industry, international conservation NGOs, and multilateral institutions—and specifically to discuss the most pressing issues around oil extraction in fragile environments such as the Amazon Basin.

The stakeholders were represented by these three main actors, thus:

- Indigenous leadership of the Amazon Basin was represented by the *Coordinadora Indígena de la Cuenca Amazónica* (COICA), as well as the national federations, including non-members of COICA.
- Governments of the Andean countries represented by OLADE (*Organizacion Latinoamericana de Energia*) and their ministries of energy and the environment, as well as state institutions responsible for indigenous affairs
- The oil and gas industry in the Amazon Basin (both private and public companies) were represented by ARPEL (*Asociacion Regional de Companias del Sector de Petroleo, Gas y Biocombustibles en America Latina y el Caribe*)

Through dialogue, EAP aims to promote development of the hydrocarbon industry in a sustainable manner in the Amazon region and to raise the social and environmental standards of the oil and gas industry in this vulnerable context, through the creation of trust and mutual understanding among governments, industry, and indigenous peoples.

Through meetings and conferences, EAP became a unique platform to jointly review environmental and social impacts, difficulties and challenges in communication, and analyze their various experiences from the oil and gas industry. By sharing participants' experiences and enabling comparative analysis, the parties could find concrete solutions at the regional level which can be applied at the national level by each of the participating nations. The EAP also became the only forum in the Andean region that brought together governments, industry, and indigenous peoples at the supranational level and allows them freedom to discuss, agree, and plan specific activities to improve the social accountability of the oil and gas sector.

The decision-making process lies with these actors, while the role of the Bank was to promote the process, provide technical guidance, act as a facilitator, and raise financial support through international development assistance. The EAP Program began by creating a small committee with representatives from the three actors, responsible for preparing a draft working agenda. This was to be discussed and approved annually at a plenary session to be held in rotating host countries. The EAP is based on a number of principles to which all parties must agree. These include recognition of each party's interests and respect for their internal decision-making processes; ensuring effective representation of each party; voluntary participation, and consensus-driven selection of issues, work dynamics and processes. The EAP framework has provided an innovative modus operandi by which the three main actors in the hydrocarbon sector have identified the nine most critical issues they face collectively, in terms of the interests, visions, and concerns of each party which they need to collectively address through the tripartite dialogue which includes:

1. Institutional strengthening of the involved parties;
2. Mechanisms and processes for consultation with indigenous peoples;
3. Participatory monitoring of the industry's environmental and social impacts;
4. Oil rent distribution and benefit sharing;
5. Compensation principles and methodologies;
6. Management and resolution of conflicts;
7. Regulation of protected areas regarding oil and gas activities;
8. Sustainable development at the local level;
9. Treatment and solution of environmental liabilities.

The EAP tripartite dialogue which spanned a total of nine years in an 11-year was in two phases (1999–2003 and 2007–10), and has recognized as one of the longest standing of the Bank's programs in terms of engagement with high-level indigenous leadership. The first EAP phase, 1999-2003, ended because of a request from COICA to make the Program a space for negotiating national agreements. This was not possible because the regional nature of the tripartite dialogues lacked national authority. The second phase was launched after internal elections to the COICA board, who requested that the EAP be reopened.

During its two phases, the Program managed to complete the first four tasks successfully and the remainder remained unaddressed for lack of sufficient funding from donors. During the second phase, at COICA's request, the EAP Program itself was subjected to consultation with indigenous leaders representing local federations. This was an unexpected development, but fortunately government and industry actors agreed, as long as they were allowed to participate in the consultation process. Consultations took place in Colombia and Peru in 2010, including representatives of local indigenous federations for the first time, leading the three actors to reach agreement stipulating that:

-   The tripartite dialogues must have the required representation at each instance (local, national, regional) and whenever possible there should be continuity in participants.
-   Dialogues at all levels should reach agreement by consensus, which should be binding, including compliance schedules that would enable parties to meet their responsibilities.
-   The tripartite dialogues should continue, including precise recommendations reached by consensus on the way these should be prepared and conducted.

Despite the difficulty involved in reaching binding consensus on all agreements, the inclusion of local indigenous representatives made the agreed-upon framework a significant step forward in oil-industry accountability to the populations it affects. The Program identified the priority issues mentioned above, through analytical and comparative papers and publications (including publications from the Bank's Energy Sector Management Assistance Program (ESMAP), covering social and environmental aspects). It also delivered workshops and capacity-building programs on:

- Environmental and social regulations of oil and gas operations in sensitive areas of the Sub-Andean Basin.
- Project cycles in the hydrocarbon sector (training for indigenous leaders).
- A study on oil rent distribution in Bolivia, Colombia, Ecuador and Peru.
- Analysis of the free, prior and informed consultation processes with indigenous peoples in Bolivia, Colombia, Ecuador and Peru.
- Proposals for the participatory monitoring of the oil and gas industry in Bolivia, Colombia, Ecuador and Peru.

The EAP was recognized as a mechanism to prevent and resolve conflicts in the extractive jurisdictions. Also, the recommendations emerging from improved consultation and participation of indigenous peoples became and remained industry standards for the Amazon Basin. However, only the Colombian and Peruvian tripartite dialogues are being shaped according to EAP principles and dynamics but all in all, the EAP has been helpful in creating references and standards regarding:

- The consultation process with indigenous peoples in the region, through analysis of existing consultation processes and countries' legal frameworks.
- Participatory monitoring of the oil and gas industry, through the analysis of actual participatory monitoring and countries' legal frameworks.
- The creation of trust between the three main actors, through 11 years of open dialogue, workshops and regional tripartite conferences and meetings.
- Building understanding among indigenous leadership about how the oil industry operates, through workshops and scholarships.
- The establishment of procedures and mechanisms for the realization of the national tripartite dialogues, in particular in Colombia and Peru.

3. THE WORLD BANK ROLE IN THE EXTRACTIVE INDUSTRIES TRANSPARENCY INITIATIVE

It is believed that about 3.5 billion people live in countries rich in oil, gas, and solid minerals. These natural resources are usually the property of the State and when extracted and revenues generated from the exploitation are well managed, these natural resources have tremendous opportunity to enhance economic growth and social development. But if poorly managed it can lead to corruption and conflicts. Some countries that have experienced natural resource-related conflicts include Angola, Liberia, Sierra Leone, Timor Leste, Iraq, Kuwait, and Afghanistan, among others. Therefore in the spirit of '*social contract*', citizens reserve the right to know how these resources benefit them.

With good governance, the exploitation of these resources can generate large revenues to foster economic growth and reduce poverty. However, when governance is weak, such resource revenues may result in corruption and potentially even conflict, as groups fight over access to these resources or the associated revenue. This can make poverty deeper and more widespread than before the resources were discovered.

In reality, this sector has over the years been subjected to secrecy and opacity. Due to this secrecy and opacity, many government officials entrusted to manage the financial resources from the extractive sector have corruptly converted same to private money and hidden away in different parts of the world. This practice of converting common wealth into private wealth has become a source of different forms of conflicts in the developing countries where these natural resources extractions take place. Therefore to reverse the secrecy, opacity and corrupt conversion of public resources, there must be more openness around how a country manages its natural resource wealth to ensure that these resources benefit all citizens and reduce conflicts.

In response to these challenges, the Publish What You Pay (PWYP) campaign was launched in June 2002 by George Soros, chairman of the Open Society Institute and Global Witness among other NGOs. The small founding coalition of NGOs was soon joined by others such as Catholic Relief Services, Human Rights Watch, Partnership Africa Canada, Pax Christi Netherlands and Secours Catholique/CARITAS France, along with an increasing number of groups from developing countries. The coalition has grown extensively since the campaign's launch and continues to expand worldwide as more than 300 organizations from several countries have signed up to the PWYP appeal.

The Publish What You Pay campaign (PWYP) aims to ensure transparency over the payment, receipt and management of oil, gas and mining revenues in resource-rich developing countries so that citizens can hold their government to account for the expenditure of this important income. Promoting resource revenue transparency is consistent with international objectives on improving governance and the budget process, enhancing corporate social responsibility, and alleviating poverty. PWYP calls for annual net figures of these payments to be disclosed by all extractive companies (both multinational and state-owned) for every country of operation. PWYP is actively engaged internationally and locally in the Extractive Industries Transparency Initiative (EITI) as one important vehicle to increase transparency in the extractive sector.

The mismanagement of the extractive industries generated-revenues is the major cause of the many conflicts in resource-rich developing countries. These conflicts range from mere civil protests organized to call on the attention of the government on the need to ensure prudent devolution and investment of funds from the extractive sector to arms confrontation against the extractive companies and government interests by militants such that operated in the Niger Delta in Nigeria, to wars such was fought in part over access to oil in the Middle East and wars fueled by "blood diamonds" in West Africa.

The continued concern about increasing level of resource mismanagement, the increasing rate of poverty in the resource-rich developing countries and how these are manifesting to conflicts, propelled the leadership of Publish What of Pay Coalition under George Soros to approach the British Government to present the issue to the leadership of the world community during the World Summit on Sustainable Development in South Africa held in 2002. The British Government bought into PWYP Coalition argument and tabled the issue at the Summit. The issue when tabled before the world leaders received tremendous support and this ushered in the now well celebrated Extractive Industries Transparency Initiative (EITI), which was launched in 2002 at the World Summit on Sustainable Development in South Africa.

From inception, the EITI has been a global coalition of governments, extractive companies and civil society working together to improve accountability in the management of revenues from natural resources sector. The EITI aims to strengthen governance by improving transparency and accountability in the extractives sector. It is a global standard that promotes revenue transparency and has a robust yet flexible methodology for monitoring and reconciling company payments and government revenues at the country level. Each participating country discloses these payments in an annual EITI Report that supposed to allow citizens to see for themselves how much revenue their government is receiving from their country's natural resources. The 'reconciliation' process consists of an independent auditor collecting, for a given period, government data on payments from extractive companies and company data on payments to government. Theoretically, these data should match. Where they do not, the discrepancy will have to be resolved by either government or the companies explaining payments differentials. Where no convincing explanation to a discrepancy between receipts and payments can be found, civil society will demand answers with respect to what happened to the relevant funds.

The World Bank endorsed the EITI initiative in 2003 and has since then, provided tangible governance improvements in the implementing countries. It works with multiple stakeholders—a coalition of governments, extractive companies, investors, international organizations and Civil Society Organizations (CSOs)—to manage a process of publication and verification of company payments and government revenues from oil, gas and mining.

Each implementing country creates its own EITI process, which is overseen by participants from the government, companies, and national civil society while the international EITI Board and the International Secretariat serve as the custodians of the EITI methodology at global level. The implementation of the EITI in these countries is a way of tackling the very causes of resource-related conflicts which further manifests into other social challenges.

EITI helps the civil society in their monitoring of revenues generated from natural resources sector and potently attempts to hinder corruption, thus reducing the risk of new resource conflicts and post-conflict countries relapsing into fighting, thus contributing to wider peace and reconciliation processes in resource conflict prevention and management.

Apart from contributing its fund, the World Bank Group source funds from other donor agencies which they deploy to providing technical and financial assistance to almost all countries implementing EITI globally. There is a Multi-Donor Trust Fund (MDTF) administered by the Bank which is used to support EITI implementing countries. The World Bank also makes experts available to advise and serve as consultants to governments, sharing international best practice, and providing grants to governments to help support EITI implementation.

The World Bank assists the CSOs to engage in the EITI process by financing and delivering training—with partners—both on the general effectiveness of CSO work and on the specifics of financial transparency and the extractive industry. In fact, the EITI initiative has proven to be a veritable mechanism for dialogue and as a springboard for concrete actions on transparency, for effective management of natural resources wealth to avoid or mitigate social conflicts in resource-rich developing countries and for good governance.

With the assistance of the World Bank the EITI principles are devolving to sub national level through the application of the EITI principles and criteria to local government revenue flows. This has become necessary in view of the contentious revenue sharing issues among the federal, the state and local government especially in such countries where natural resources exploitations are concentrated in a region, like in Nigeria, DRC and Sierra

Leone among other countries. This is crucial for stability, as regional representatives from the resource rich areas often believe their region does not get a fair share of "its" resource wealth, leading to grievances or even violent independence movements.

As of November 2013, there were 41 EITI implementing countries, including the recently approved candidate applications from the Philippines and Honduras. In addition, over 80 companies support the EITI—international, state-owned as well as extractive and non-extractive companies. Over 90 institutional investors with total assets of more than US$19 trillion support the EITI. Of the 41 EITI countries, 25 countries were adjudged EITI compliant. Through the EITI, resource-rich countries have increased their understanding of their fiscal position and thus gained a clearer picture of the risks to their fiscal prospects[21].

Today, due to the involvement and support of the World Bank in the EITI, many partner organizations such as the Open Society Initiative for Western Africa (OSIWA), PACT, SwissAid, Catholic Relief Services (CRS), Oxfam, Revenue Watch Institute (RWI), the Budget Project Institute, and the Friedrich Ebert Stiftung (FES), among others are now playing critical roles with both the host states and the civil society organizations members of PWYP to ensure sustainable development of the extractive sector. They support in reinforcing capacities, institutionalization, communication and training; technical and financial support, budget monitoring, promoting economic justice and managing extractive resources in a sustainable way.

4. HOW THE UNITED NATIONS IS CONTRIBUTING TO SUSTAINABLE EXTRACTIVE SECTOR MANAGEMENT

The protection of international human rights standards has traditionally fell under the exclusive charge of governments, with the responsibilities to regulate relations between the State and individuals and groups.

The process of globalization over the past decades have seen non- state actors such as transnational corporations and other businesses play significant part internationally, nationally and at local levels by giving rise to a debate about the roles and responsibilities of such actors with regard to human rights. With the increased role of corporate actors, nationally and internationally, the issue of business' impact on the enjoyment of human rights has been placed on the agenda of the United Nations.

---

[21] World Bank (2013) Building on Progress to Implement the EITI Standard, EITI MDTF 2013 ANNUAL REPORT No. 83538 Pg. 9.

The increased understanding of the world community on problems associated with the extractive sector in developing countries has elevated the needs for critical steps to be taken to turn the resource curse in this sector into resource blessing.

With this expansion phenomenon, the United Nations human rights mechanism started to scope business' human rights responsibilities and exploring ways for corporate actors to be held accountable for the impacts of their activities on human rights. As a result of this process, there is now greater clarity about the respective roles and responsibilities of governments and business with regard to protection and respect for human rights.

This clarity has amplified social awareness of businesses' impacts on human rights and also attracted the attention of the United Nations to the linkage between business and human rights thus creating an opportunity for the formation of global policy agenda in the 1990s.

There is now growing awareness that over the past few decades, extractive companies have increasingly come under serious scrutiny from other key stakeholders because of the persistent negative impacts of their activities in both the host and impacted communities. The world community has also through one convention or another or resolutions and strategic frameworks tried to assist extractive companies in meeting the dual needs of their social responsibilities to host and impacted communities on one hand and the extractive companies' quest for profit on investment, satisfaction of shareholders and business sustainability on the other hand.

For instance, the Free, Prior and Informed Consent mechanism of 2007 United Nations Declaration on the Rights of Indigenous Peoples and the 1989 International Labor Organization's Convention 169; the adopted United Nations Guiding Principles on Business and Human Rights; the Sustainable Development Goals; the Equator Principles and Global Compact framework among others now make the respect for human rights a major obligation of business organizations no matter how large or small, to seek out ways and strategies to operate their extractive businesses without compromising human rights in their center of operations.

With the above frameworks, human rights-based approach to sustainable extractive sector management which is best described as a conceptual and procedural framework of ensuring the promotion and protection of human rights in projects planning and execution have become a necessity for successful and sustainable business operations for the extractive companies as well as the basis to:

1. Position human rights and its principles as the core component of extractive companies development actions;

2. Demand accountability and transparency by host and impacted communities;'
3. Foster empowerment and capacity building of host and impacted communities to, inter alia, hold the extractive companies to account;
4. Ensure that the meaningful participation of host and impacted communities in development processes and planned interventions is recognized as an intrinsic right, not simply as best practice publicity stunt; and
5. Ensure the non-discriminatory engagement of host and impacted communities and the prioritization of the needs and rights of especially the vulnerable or marginalized individuals or groups such women, the elderly, the children and youths, and the minorities and Indigenous peoples.

Further, all social, economic, cultural, civil, political and environmental rights are now regarded as human rights and there is genuine understanding that the host and impacted communities have legal entitlements. With these common knowledge and the quest to ensure maximum benefits of investments for investors and the greater society, there is need for the extractive companies and government to:

1. Positively enhance the social, economic, environmental and human impacts of their extractive activities;
2. Adapt and expand project infrastructure to ensure it can also: Service legitimate local community needs, Provide funding and enhance technical capacity to support local host and impacted communities sustainable development plans, Ensure that in case of resettlement resulting from the business activities of extractive companies, the post-resettlement outcome will restore livelihoods of the negatively affected persons; Enhance genuine commitment to local content vehicles to make it possible for local enterprises in the local host and impacted communities to supply goods and services, Enhance the process of identifying and engagement of affected communities in participatory processes of identification, assessment and management of social impacts of projects;
3. Enhance the benefits of projects to impacted communities to achieve better outcomes for affected communities and extractive companies projects promoters;
4. Reduce the level of business promoters risk exposure;
5. Become compliant with international standards and good corporate practices; and
6. Address significant human rights issues associated with extractive companies' projects and products.

For this reason, there has been formal and informal institutional mechanisms put in place to address the problems and these challenges. In this regards therefore, four of such mechanisms are addressed here to wit:

4.1 UN Global Compact.

4.2 UN Guiding Principles on Business and Human Rights.

4.3 The United Nations adopted Sustainable Development Goals.

4.4 The U N Human Rights Council Resolution on Environmental and Human Rights Abuses in Nigeria.

4.1 The UN Global Compact

This global policy agenda gave rise to the UN Global Compact framework which became the leading global voluntary initiative for corporate social responsibility that also addresses the issue of business and human rights. This initiative aimed to bring international businesses in alignment with the Universal Declaration of Human Rights.

The UN Global Compact was launched in July 2000 not as a regulatory mechanism but as a nebulous intentional flexible channel through which policy dialogues, learning, local networks and projects are facilitated and encouraged. It is just a platform for the networking of key stakeholders in sustainable development such as governments, companies, the labor organizations and civil society organizations. The UN Global Compact however recognizes that companies' declaration of support for the principles does not imply that these companies have fulfilled the principles of the Compact.

The United Nations Global Compact is a framework to guide all businesses regardless of size, complexity or location and is being implemented in about 160 countries around the world by over 8,000 companies within the realms of human rights, labor, Environment and Anti-corruption. It is intended to transform the world of business and social responsibilities through its Ten (10) Principles.

The Ten Principles of UN Global Compact

The UN Global Compact establishes 10 principles that corporations must address in their quest for sustainable development which include:

A. Human Rights

Businesses should:

- Principle 1: Support and respect the protection of internationally proclaimed human rights; and
- Principle 2: Make sure that they are not complicit in human rights abuses.

B. Labor Standards

Businesses should uphold:

- Principle 3: The freedom of association and the effective recognition of the right to collective bargaining;
- Principle 4: The elimination of all forms of forced and compulsory labor;
- Principle 5: The effective abolition of child labor; and
- Principle 6: The elimination of discrimination in employment and occupation.

C. Environment

Businesses should:

- Principle 7: support a precautionary approach to environmental challenges;
- Principle 8: undertake initiatives to promote environmental responsibility; and
- Principle 9: encourage the development and diffusion of environmentally friendly technologies.

Anti-Corruption

- Principle 10: Businesses should work against corruption in all its forms, including extortion and bribery.

Requirements for Effective Implementation of the UN Global Compact

The UN Global Compact Management Model is comprised of six steps with each step requiring one or more suggested activities and areas of focus. The management model includes the following steps:

Step 1: Business Leadership Commitment to mainstream the UN Global Compact Principles into strategies and principles and to take actions in support of broader UN goals, in a transparent way.

Step 2: Company assessment of the risks, opportunities and impacts across UN Global Compact issue areas.

Step 3: Company's definition of goals, strategies and policies.

Step 4: Company's implementation of strategies and policies through the company and across company's value chain.

Step 5: Company measures and monitors impacts and progress towards goals.

Step 6: Company communicates progress and strategies and engages other stakeholders for continuous process improvement.

Effectiveness of the UN Global Compact

The UN Global Compact has been adjudged as ineffective by the civil society organizations, leading UN project experts and other individuals and organizations as it does not have any effective monitoring and enforcement provisions, therefore, cannot be used to hold corporations accountable. In fact many critics regard the Global Compact as a public relations stunt which can be used as a pretext and to oppose any binding international regulation on corporate accountability as well as an entry-point to increase corporate influence on the policy discourse and the development strategies of the United Nations. The Global Compact has been alleged to have admitted companies with dubious humanitarian and environmental records in contrast with the principles demanded by the Compact.

The UN Global Compact has been also alleged to have failed in all the areas it is meant to intervene. For instance, in the area of human rights, host and impacted communities are abused by corporations every second of the day especially in the developing world where the extractive companies operate. There is still abundant evidence of child labor in the mining industries, as well as forced labor especially by state security forces who are engaged by the corporations to clear right-of-way, carry equipment and provide security among others.

In the area of environmental pollution, there is the continued cases of oil spillage, gas flaring, dumping of toxic effluents on water channels, land and burning them too. The poor people being those who are suffering the consequences of the problems more than those that created them.

Corruption is ever-present and the voluntary UN Global Compact framework has done very little to mitigate it. Since signing this UN Global Compact, many businesses have been allegedly linked to corruption across the globe.

In summary, the UN Global Compact though a nice framework, yet it lacks what is needed for sustainable business operations and sustainable development.

4.2. The UN Guiding Principles on Business and Human Rights

In further recognition of the nexus between business and human rights, the United Nations through the then Commission on Human Rights appointed an expert subsidiary body to draft what was called the *Norms on Transnational Corporations and Other Business Enterprises* aimed at imposing on companies, directly under international law, the same range of human rights duties that States have accepted for themselves under treaties they have ratified which include "*to promote, secure the fulfilment of, respect, ensure respect of and protect human rights*".

This move created acrimonious relationship between the business community and human rights advocacy groups while the Governments played the "possum". As a result of Governments inability to fully come out with maximum support for the initiative, the then Commission on Human Rights dropped the implementation of the draft proposal. Instead, in furtherance of the need to effectively respond to the issue of business and human rights, in 2005 the UN established a mandate for a Special Representative of the Secretary-General "on the issue of human rights and transnational corporations and other business enterprises" to undertake a new process, and requested the Secretary-General to appoint the mandate holder.

To fulfill the mandate, Professor John Ruggie of Kennedy School of Government at Harvard University, was appointed the Special Representative of the Secretary-General on the issue of human rights and transnational corporations and other business enterprises, in 2005.

The Human Rights Council in resolution 8/7, the High Commissioner for Human Rights convened a two-day consultation in October 2009 on the issue of human rights and transnational corporations and other business enterprises. The meeting brought together the Special Representative, business representatives and all relevant stakeholders, including non-governmental organizations and representatives to discuss ways and means to ensure successful implementation of the "Protect, Respect, Remedy" Framework on business and human rights.

In 2011 the report of the Special Representative of the Secretary-General on the issue of human rights and transnational corporations and other business enterprises, was adopted by the UN Human Rights Council as the "*Guiding Principles on Human Rights and Business Implementing the United Nations 'Protect, Respect and Remedy' Framework*". This UN Guiding Principles on Business and Human Rights *institutes* business responsibilities for the promotion and protection of all human rights: *civil, political, economic, social and cultural rights, including the right to development.*

The UN has now put in place the "Protect, Respect and Remedy" Framework on business and human rights, which was elaborated by the Special Representative of the UN Secretary-General on the issue of human rights and transnational corporations and other business enterprises, building on major research and extensive consultations with all relevant stakeholders, including States, civil society and the business community.

On 16 June 2011, the UN Human Rights Council endorsed the Guiding Principles on Business and Human Rights for implementing the UN "Protect, Respect and Remedy" Framework, providing – for the first time – a global standard for preventing and addressing the risk of adverse impacts on human rights linked to business activity.

The framework of this UN Guiding Principles on Business and Human Rights rests on three pillars as set forth below:

1. The first is the State duty to protect against human rights abuses by third parties, including business enterprises, through appropriate policies, regulation, and adjudication.
2. The second is the corporate responsibility to respect human rights, which means that business enterprises should act with due diligence to avoid infringing on the rights of others and to address adverse impacts of the business activities with which they are involved.
3. The third is the need for greater access by victims of human rights abuse by businesses to effective remedy, both judicial and non-judicial.

These three pillars of business and human rights are critical components of widely acknowledged inter-related and dynamic system of preventative and remedial actions which include:

I. The State duty to protect human rights

The State has the duty to protect human rights because it lies at the very core of the international human rights regime;

States must protect against human rights abuse within their territory and/or jurisdiction by third parties, including business enterprises. This requires taking appropriate steps to prevent, investigate, punish and redress such abuse through effective policies, legislation, regulations and adjudication.

II. The corporate responsibility to respect human rights

The corporate responsibility to respect human rights because it is the basic expectation society has of business in relation to human rights.

Business enterprises should respect human rights. This means that they should avoid infringing on the human rights of others and should address adverse human rights impacts with which they are involved.

III. Access to remedy

Access to remedy is provided for because even the most concerted efforts cannot prevent all abuses.

As part of their duty to protect against business-related human rights abuse, States must take appropriate steps to ensure, through judicial, administrative, legislative or other appropriate means, that when such abuses occur within their territory and/or jurisdiction those affected have access to effective remedy.

In addition to the Human Rights Council, individual Governments, business enterprises and associations, civil society and workers' organizations, national human rights institutions, and investors have endorsed this Framework. Equally, the Framework has been drawn upon by multilateral institutions such as the International Organization for Standardization and the Organization for Economic Cooperation and Development in developing their own initiatives in the business and human rights domain. Other United Nations special procedures have also extensively invoked the Framework.

However, based on the available experience and documented facts, in light of what is trending in the extractive sector reported in this book and elsewhere, this framework is unarguably new therefore must be effectively implemented bearing in mind all its key provisions, to mitigate business abuse of human rights especially in the extractive sector.

4.3 The United Nations Sustainable Development Goals (SDGs)

April 22 of every year has been designated the "World Earths day" which symbolizes the anniversary of the birth of the modern environmental movement started in the United States of America and European countries especially, dating back to the 1970s.

The need to protect earth's natural resources from the extractive sector amid the climate change threats resulting from continued deforestation, excavations and pollution of the natural environment by the extractive companies continues with increasing urgency especially with the United Nations Food and Agriculture Organization (FAO) enlightenments and disclosures that close to 1.6 billion people or more than 25 per cent of the world's population are dependent on forest resources for their livelihoods.

In September 2015, the United Nations agreed on a set of 17 development goals known as Sustainable Development Goals (SDGs) as the global Agenda for transforming the world by the year 2030. The 17 Sustainable Development Goals with their 169 targets which seek to balance the three dimensions of sustainable development: the economic, social and environmental has elevated the SDGs to the level of an integrated approach to Sustainable Development which is necessary for the extractive sector to imbibe to ensure their successful operations and protection of the earth and its people.

The Sustainable Development Goals (SDGs), are the universal set of goals, targets and indicators that UN members will use to frame their policies over the next 15 years. They are now a centerpiece of international efforts to chart a more responsible course and provide a road map to the universally articulated and accepted sustainable development.

One critical importance of the SDGs is that each of the 17 SDGs identifies a subject of importance. Specifically, goals 1 and 2 deal with poverty eradication and food security respectively while other goals deal with such issues as gender equality, climate, etc., all of which are critical cross-cutting issues in the extractive sector and in determining whether the extractive sector will honestly contribute to sustainable development especially in their areas of operation.

The provision of "Sustainability Indicators," in the SDGs will help the stakeholders in the extractive sector to evaluate and decide the opportunities and challenges of the extractive sector in contributing to reduction in climate change, reduction in poverty in the extractive jurisdictions, reduction in human rights abuses and contributing to the practical achievement of sustainable development.

One major importance of integration of economic, social and environmental factors in the SDGs is the need for these developmental components not to be pursued independently but through more effective inter-linkages with other entities.

The SDGs integration approach presupposes that the previously held belief of disaggregation has not worked and could alienate distinct interest groups instead of facilitating collaborative achievement of other goals like ensuring viable livelihoods for local populations, environmental protection and respect to human rights.

The United Nations resolve to shift focus towards more integrated indicators using the SDGs, really holds a lot of promise and a broad relevance that support the view that different aspects of each system are tightly interlinked, which means that striving to improve one indicator may positively affect others. The clear lesson stemming from this understanding will effectively improve the management of the extractive sector, if judiciously pursued.

4.4. The U N Human Rights Council Resolution on Environmental and Human Rights Abuses in Nigeria

The United Nations Human Rights Council sitting in Geneva in 2014 passed a resolution to hold transnational companies, especially in the oil and gas sector, accountable for environmental and human rights abuses in Nigeria, and anywhere in the world that they operate.

The resolution jointly sponsored by Ecuador and South Africa is the result of many years of pressure by communities, local and international NGOs and social movements across the world demanding change to save people and the environment. This resolution for a uniform legally binding instrument was supported by over 610 organizations, 400 individuals, and 95 countries while 13 Nation states abstained. The successful passing of this resolution demystified the sense of corporate imprisonment of the UN.

This resolution represents a noteworthy and extraordinary triumph for the defenders of the environment, human rights and sustainable livelihoods from the violations of big business. This victory means expansion and advancement of access to justice, right to protest and protection of environmental defenders, reparation and remediation of damaged environment and livelihoods, and criminal liability for corporate offenders.

This monumental victory means that the decades of mobilising and resisting the negative consequences of the activities of the extractive industries by local, national and international Civil Society Organizations working to ensure environmental and social justice were never in vain. It is a demonstration of the significance of people's power.

This uniform binding mechanism will ensure that Shell and other oil companies in Nigeria will apply the same standards deployed in Europe and America in their operations in Nigeria and elsewhere.

Further, it would bring to an end corporate impunity that is undermining national governments and institutions such as Nigeria.

5. CHEVRON'S ACTIVITIES TOWARDS SUSTAINABLE DEVELOPMENT IN NIGER DELTA

There is prevalent human rights abuses and environmental pollution in the Niger Delta from the extractive activities of the oil and gas companies operating in the area. The waters are polluted from oil spillage, killing the fishes and other aquatic mammals; the lands are polluted from dumping of industrial wastes and oil spills and the air is polluted through gas flaring that drive away animals deeper into bushes and acid rain. In addition, there is a high level of poverty among the majority of the Niger Delta population.

Royal Dutch/Shell, Chevron, AGIP, ExxonMobil, ELF, Texaco have been blamed for the fate that has befallen the Niger Delta. Although many of these companies have one way or another tried albeit their inconsequentialities to ameliorate the consequences of their actions, but the success rate have been negligible. However, this section of the book will zero-in on the steps Chevron International is currently taking towards reversing its approaches to community empowerment in its jurisdiction of operation.

Chevron which is a San Francisco, California based international oil company has had its oil and gas operations in Nigeria, dating back to the early 1960s. Since coming into Nigeria the company has exploited oil and gas resources in the Niger Delta that houses the most lucrative oil blocks and oil fields where oil production in the country is currently taking place.

Chevron Oil Company has had successful operation in Nigeria from its revenue profile and has contributed immensely to economic development of the country in many meaningful ways that include payment of tax and royalties to the federal government, payment of employees' income tax to the state governments, contribution to the development of host communities through provision of educational, health, water, electricity and transportation infrastructure; agricultural development, offering scholarships and employment opportunities to the citizens of Nigeria, within its Corporate Social Responsibilities framework boundaries.

Despite the above contributions, Chevron Nigeria Limited was able to acknowledge the fact that its corporate social responsibility was not leading to sustainable development in the Niger Delta as it declared its company's corporate social responsibility framework thus:

"The company's new approach is focused on sustainability. This sustainable development approach is informed by the lessons we have learnt and the unsatisfactory results we got from our previous approach. We have been involved in community development in the Niger Delta for many years… since the start of our operations in the early 1960s. But when we try to analyze what we have got to show for our long running engagement with the communities, what we see are school buildings, hospitals, jetties, town halls, school libraries, blocks of classrooms, market stalls and so on. The question can then be asked: All these projects what do they add up to? Do they actually cause development to happen in the communities? Not quite as well as we would like to see happen. It is true we have caused some level of improvement in the life of the community people by providing good water, good schools and so on. These are good things but talking about causing development to happen, not really…. We needed a new strategy that will drive the engine of growth in the communities and that strategy is what we call sustainable development. It ensures development in the communities that will outlive the present generation, improve the life of the people and stand a good chance of guaranteeing the future of their children. It should not depend on Chevron to survive. It should be self-sustaining and add value to people by helping them to build capacity" (CR Magazine, Vol. 1 Number 3, Chevron Texaco Nigeria, 2004). In recognizing the need for a change in the way Chevron Texaco relates to the communities, it has decided to change its strategic approach.[22]

This change is coming with the establishment of the **Foundation for Partnership Initiatives in the Niger Delta (PIND)**, which is a non-profit foundation of Chevron's initiative and administered by **Niger Delta Partnership Initiative (NDPI)** based in Washington D.C., to provide support for socio-economic development programs in the Niger Delta (ND) region where it operates.

The PIND initiative seeks to create dynamic, multi-stakeholder partnerships that take full advantage of the synergies of involving other donors from the public and private sector and diverse organizations and interests in development support programs and activities, which empower communities to achieve a peaceful and enabling

---

[22] See Thisday Newspaper. Tuesday May 4, 2005 vol.11. No. 3664 pg 32-33; Vanguard Newspaper Tuesday May 10,2005 vol. 21: No 5674 pgs. 1 and 13.

environment for overall equitable economic growth in the Niger Delta. The PIND project is focused on four major activity areas which include:

4.1. Economic Development-Facilitates opportunities for pro-poor market development.

4.2 Capacity Building-Build the service delivery and engagement capacity of government, civil society and communities.

4.3 Analysis & Advocacy-Improve analysis and understanding of systemic constraints to growth.

4.4 Peace Building-Strengthen integrated conflict resolution that enables economic growth.

The specific activities undertaken by the PIND project which is expected to create enabling and peaceful environment for economic growth, increased equitable economic growth and increased income and employment in the Niger Delta under the above strategic objectives are briefly discussed under their various subheadings.

4.1 Economic Development Objectives

The objectives of PIND economic development are to:

1. Enhance and promote economic development services (EDS) in the Niger Delta;
2. Establish and maintain training, research, ICT and office facilities to support EDS activities and projects throughout the Niger Delta;
3. Build capacity and network of local partners in the Niger Delta to support EDS;
4. Establish and maintain asset base and field office that will enable PIND to provide long term support for development in the Niger Delta;
5. Focus on achieving broad-based economic growth to increase incomes and employment in the Niger Delta.

Economic Development Services

The areas of focused economic development services are through value chain development in the following agricultural sectors:

- Aquaculture
- Cassava
- Palm Oil
- Poultry

Enterprise Development:

- SME Development: This involves active collaboration with GroFin.
- Business Linkages: This involves building the capacity of local businesses to meet the needs of multinationals in areas of quality of products and services, and competiveness.

Program Development Process

It is envisioned that development Program can be achieved by:

- Identifying opportunities based on value chain or other analysis and potential for pro-poor growth;
- Developing an understanding of the binding constraints that need to be addressed to reach the potential;
- Implementation of pilot project/ activities to test the market driven solutions to realize the market growth potential, starting with a growth node;
- Successful activities will be used as examples to expand and promote wider adoption by other implementing partners and development actors.

4.2 Capacity Building-Build the service delivery and engagement capacity of government, civil society and communities.

Capacity Building: Program Objective

The established objective here is to build the capacity of government, civil society and communities to engage in effective service delivery, economic development and peace-building activities in communities and the Niger Delta region

Capacity Building Program: Projects Completed

According to PIND, it has through the Advocacy Awareness and Civic Empowerment (ADVANCE) provided small grants to some CSOs for activities that promote enabling environment for economic growth in the region, training 1,573 male and 1,496 female community leaders to take actions that have improved service delivery and provisions of economic infrastructure. This though is subject to verification to authenticate the veracity of this success story.

Also, according to PIND, it has through the Local Capacity Building Project (LCBP) facilitated Local Government/ community engagement inclusive process for identifying, developing and implementing infrastructural projects.

In addition, according to PIND, the Social Sector Investment Action Plan for the Niger Delta with United Nations Development Program and the Niger Delta Action plan have been developed and disseminated to Stakeholders and PIND as a member of the National Council on Niger Delta will implement the Plan.

Further, according to PIND, 56 Civil Society Organization's Program and Finance officers (34 males, 22 females) have completed CAPABLE training. The newly acquired skills according to PIND are helping them to transform their organizations with emphasis on the development of Administrative and Finance policy manuals, e-filing system, enhanced Internal Control System and documentation, proposals, newsletters, etc.

Furthermore, PIND claims that 36 youths (23 males and 13 females) have completed training in leadership, entrepreneurship and computer/communication skills as the first phase of the Niger Delta Youth Leadership capacity building project.

4.3 Analysis and Advocacy: Improve analysis and understanding of systemic constraints to growth.

The other strategic area of PIND's activities in the Niger Delta is *analysis and advocacy*. Through this focus strategy PIND says it will collect, analyze and disseminate information relevant to economic growth. The organization also indicates it will influence the public sector, the private sector and the communities to use evidence based research findings for decision making; identify new models and projects leading to economic growth, increase capacity and reduce conflicts.

4.4 Peace Building-Strengthen integrated conflict resolution that enables economic growth.

The next strategic area of PIND's activities in the Niger Delta as approved by NDPI based on the concept note submitted by PIND is peace-building in the 9 Niger Delta states through the establishment of the *Partners for Peace (P4P) in the Niger Delta*. The P4P initiative is a network of non-governmental, non-profit making and non-political network of individuals, civil society organizations, governmental units, businesses and community leaders with varied experience established to strengthen conflict resolution initiatives and build peace that will facilitate and sustain economic and human development in the entire Niger Delta.

P4P network was officially launched on 29th August, 2013, in Port Harcourt, Rivers State. At this formal launching, some representatives from the nine Niger Delta states were selected and mandated to replicate the P4P network in their various states. Today, the network has been established in all the Niger Delta states.

The core activities of P4P network are to facilitate communications, information exchange and networking to increase capacity, increase effectiveness and identify, recognize and amplify the voices of peace actors and initiatives.

The P4P has held consultation in the 9 Niger Delta states of Abia, Akwa-Ibom, Bayelsa, Cross River, Delta, Edo, Imo, Ondo and Rivers with all the nine states having a chapter of P4P Network in place and the dynamics of conflict in each state of the Niger Delta has been known and appropriate intervention being planned. Analysis shows that conflict in all the Niger Delta states are NOT the same.

Chevron's one other Contribution in Nigeria

Further, it is a public knowledge that on June 3, 2012 when the Dana Airline accident occurred in Lagos, there was no Molecular DNA testing facility in Nigeria. This was a huge challenge as the Lagos state government was said to have literally spent over Twenty Two Million Naira (N22m) to send samples of the remains of the victims of the crash overseas for identification.

Following this embarrassing situation, Medical Researchers at the Lagos University Teaching Hospital (LUTH) approached the Corporate Social Responsibility Department of Chevron Nigeria Limited for assistance in acquiring the technology for Molecular DNA testing. Chevron provided the needed fund to set up the Chevron Molecular Biology Laboratory at LUTH which was commissioned in February 2014.

With this laboratory, the Ebola disease victim who introduced the disease into Nigeria on July 20, 2014, Patrick Sawyer was successfully diagnosed, making it possible for Nigeria to contain the spread of the disease, quickly identifying those infected and isolating and treating them effectively and finally being declared by the World Health Organization just within three months of first contact as Ebola Disease Virus free country on October 20, 2014.

Had Chevron not provided such facility, humanity would just be left to imagine what could have happened in Africa's most populous nation and the entire world as the rate of transmission and infections linked to contacts from domestic and international travels could have become monumental.

CHAPTER SIX

Some Impediments To Sustainable Extractive Sector Management

The extractive sector is a veritable economic sector that can serve as the foundation from which rural poor people in the host and impacted communities and beyond where the extractive companies carryout their economic activities can overcome poverty and infrastructural development deficit. However, the extractive companies developing the natural resource in these poor communities always seem to forget that their activities especially in the developing countries where they operate are putting the world and its people at a tipping point on key social challenges relating to their extractive activities and need to radically scale up actions to circumvent their devastating consequences on the earth and its people.

Every business activity has its ultimate gold standard to reach which is that last chain link in the input through the outcome spectrum known phraseologically as "impact". In development parlance, this ultimate gold standard is referred to as socio-economic impacts. The extractive sector like every other economic sector has its socio-economic impacts which can be positive or negative, envisioned or accidental, momentary or sustainable over time.

These positive socio-economic "impacts," manifest to positive changes in assets, capabilities, opportunities, and standards of living of the people and can result from increases in educational attainment, health status and income level or decreases in hunger and the incidence of disease which ultimately manifest to sustainable development of host and impacted communities and sustainable profit for the extractive companies as the end goal and eventual measure of success.

To achieve this, the extractive sector requires a new operational system, and needs to be re-tooled so that the financial, natural, and social capital returns on investment become harmonious, balanced and sustainable.

There is great urgency therefore for the key stakeholders in the extractive sector to address the sustainable development challenges, which include poverty, societal conflicts, social incongruences and destabilizations, climate change and ecological degradation among others that are linked to extractive sector development; and the extractive companies have a key role to play in providing the assortment of technologies, innovative strategies and capacity, resources and skills that are required to develop and institutionalize the radical solutions that can tackle the key environmental and social challenges encumbering sustainable development in the host and impacted communities.

Measuring and reporting on extractive sector corporate performance are essential components of the drive needed to achieve sustainable development in the sector. This is unarguably understandable because the extractive sector needs transformation which starts with measuring its performance. The results obtained from the measurements create room for sustainable management of the extractive sector to address an area of societal development which is their socio-economic impact on the society that has so far been neglected by the extractive companies. This is in the light of the fact that sustainable extractive sector management cannot happen without a concrete understanding of what works and what does not work and having sound measurement systems in place is fundamental to obtaining this insight.

In this light it is important to address some critical issues that unarguably impeded sustainable extractive sector management which are set forth below.

1. ACCESS TO JUSTICE

Victims of corporate human rights and environmental abuses resulting from the activities of extractive companies face severe obstacles in accessing justice and obtaining remediation through legal systems both in the host state and the home state of the business corporation such as Australia, Canada, the Democratic Republic of Congo (DRC), the European Union, France, Germany, India, Malaysia, the People's Republic of China, Russia, South Africa, the United Kingdom and the United States to mention a few.

According to Access to Judicial Remedy (A2JR) Project, "the United States, Canada, and Europe… are generally not fulfilling their obligation to ensure access to effective judicial remedies to victims of human rights violations by businesses operating outside their territory. Victims continue to face barriers that at times can completely block their access to an effective remedy. These barriers have been overcome in only some instances.

Victims of human rights violations by business, wherever the violations occur, should be entitled to full and effective access to judicial remedies. In order to provide access to justice for victims of human rights violations by the extractive industries, each nation should examine the barriers in their jurisdiction and consider the range of actions they can take to provide enabling environment for victims of corporate human rights abuses to seek redress within established national and international legal and regulatory frameworks.

The UK Government has been alleged by some international NGOs not to have lived up to its commitment to ensure that victims of corporate human rights violations and environmental abuses perpetrated overseas by UK registered corporations are able to access justice in the UK courts. However, the case of Bodo Community in the Niger Delta brought against the Royal Dutch Shell and its Subsidiary company, Shell Petroleum Development Company (SPDC) of Nigeria seemed to show a deviation from the usual character to a new facet of development.

Members of the Bodo community in Nigeria Niger Delta filed a lawsuit against Shell in London High Court on 23 March 2012, seeking compensation for two oil spills, which occurred in 2008 and 2009 in their community. The over 15,000 plaintiffs in the suit asked for compensation for losses suffered to their health, livelihoods and land, and they also asked for clean-up of the oil polluted areas. A preliminary hearing of the case to consider Shell's duty to take reasonable steps to prevent spillage from their pipelines took place from 29 April to 9 May 2014. After analyzing the evidence adduced by both the plaintiffs and defense, the judge ruled on 20 June 2014 that Shell could be held responsible for spills from their pipelines if the company fails to take reasonable measures to protect them from malfunction or from oil theft (known as "bunkering").

While the case was expected to go on trial in mid-2015, Shell agreed to a Fifty-five million Pounds (£55m) out of court settlement in January 2015. Thirty five million (£35m) Pounds of the settlement amount will be split among those negatively impacted by the spill and the balance Twenty million Pounds (£20m) will be devoted to community development projects.

Apart from this departure witnessed in the UK High Court from the norms, it remains extremely difficult for court cases against multinationals to proceed in many of the countries where the alleged abuses occurred. It is therefore essential that such cases can be brought in the home states of the companies concerned.

Governments unanimously agreed to address this problem when they endorsed the UN Guiding Principles on Business and Human Rights. As home to some of the world's largest multinationals, the UK have to lead the way in delivering on this obligation. Instead, the Government went ahead with changes to the court costs regime, which means it is now far more difficult to bring such cases in this country[23]. It is quite important the

---

[23] Lord Dholakia, Lord Phillips of Sudbury, Sir Nigel Rodley, Kirsty Brimelow QC - Chair of Bar Human Rights Committee, Nick Fluck - President of Law Society, Martyn Day - Leigh Day, Carla Ferstman - Director of REDRESS, Maura McGowan QC - Chairman of the Bar, Phil Lynch - Director, International Service for Human Rights & 12 other lawyers 02 Dec 2013.

UK Government ensures that victims of abuse perpetrated in developing countries are provided access to justice in the UK.

Increasingly in Canadian, extractive companies operating overseas and their subsidiaries have become the subject of allegations of human rights violations associated with their overseas activities, particularly when operating in developing countries. For years, liability before Canadian courts has been avoided. However, the decision of the Ontario Superior Court in Choc v. Hudbay Minerals Inc. may be the first step towards recognition that Canadian companies should be accountable for their behavior outside Canada[24].

According to the Council on Hemispheric Affairs (COHA), citing a 1997 lawsuit filed by Guyanese villagers against Cambior Corporation regarding alleged negligence surrounding a dam break disaster along the Omai River "…it is nearly impossible for foreign citizens to bring lawsuits involving egregious environmental and human rights violations in Canadian courts"…. This resulted in a catastrophe of mass contamination and fatalities." COHA also said that child labor is a "widespread problem" within the Guyanese mining industry. According to the organization, most recently, an eight-year-old child was found laboring in a gold mine near the Venezuelan border and in addition, indigenous groups lost a "crucial court case" filed against mining companies, when Guyanese High Court decided that it does not have the right to expel miners from their lands[25].

About 23,000 people live in the region surrounding the Essequibo River in Guyana, where Cambior, a Canadian-based mining company operates the Omai gold mine. The Omai Mine is wholly owned by Omai Gold Mines Limited (OGML). In August 1995, the tailings dam at the Omai Mine in Guyana failed, spilling mine tailings containing cyanide, heavy metals and other pollutants into the Essequibo River, contaminating the river that the 23,000 people living in the region depend on as a source of their livelihoods (drinking water, cooking, bathing, washing and fishing). At the time of the spill, Cambior owned 65% of the company and the balance was owned by Golden Star Resources and the Government of Guyana. A public interest group filed a class action lawsuit against Cambior in 1997 in Québec Superior Court seeking damages on behalf of the Guyanese victims of the spill.

In 2002, Cambior acquired Golden Star's interest in OGML, thereby obtaining a 95% ownership interest in the company. The Québec Superior Court dismissed the case in August 1998, on the grounds that the courts in Guyana were in a better position to hear the case. A lawsuit against Cambior was filed in Guyana, but it was dismissed by the High Court of the Supreme Court of Judicature of Guyana in 2002. A new suit was filed against

---

[24] Sonya Nigam, University of Ottawa, in Canadian Lawyer 14 Oct 2013
[25] CMC and Antigua Observer 03 Mar 2013

Cambior in 2003 in Guyana again seeking damages for the effects of the 1995 spill. In October 2006, the High Court of the Supreme Court of Judicature of Guyana ordered the dismissal of the 2003 action and ordered the plaintiffs to pay the defendants' legal costs.

Canadian mining companies seem to enjoy impunity virtually everywhere that they operate overseas especially in developing countries. The Canadian government has virtually abdicated its governance responsibility regarding the overseas activities of the mining sector as deduced form her refusing to regulate either the companies or the government agencies that support them, or to take legislative action to ensure that non-nationals whose human rights have been infringed by the extractive companies are able to seek redress in Canada.

In the US, the Alien Tort Claims Act (ATCA) represented a veritable mechanism with the most promising potential for holding MNCs to account for human rights violations in developing countries and the public interest lawyers in the United States of America were at the forefront of developing ATCA cases where MNCs are alleged to have been complicit with states in such violations. However, a majority decision of the US Second Circuit Courts of Appeals in September 2010 which was affirmed by the US Supreme Court held that customary international human rights law does not recognize the liability of corporations, and consequently that MNCs cannot be liable under ATCA has become a hope-damper in relying on ATCA[26].

The petitioners in this case are a group of Nigerian nationals residing in the United States who filed a suit in the United States District Court for the Southern District of New York, against Royal Dutch Petroleum Co., British, and Nigerian corporations under the Alien Tort Statute, 28 U. S. C. §1350, alleging that the corporations aided and abetted the Nigerian Government in committing (1) extrajudicial killings; (2) crimes against humanity; (3) torture and cruel treatment; (4) arbitrary arrest and detention; (5) violations of the rights to life, liberty, security, and association; (6) forced exile; and (7) property destruction.

The District Court dismissed the first, fifth, sixth, and seventh claims, reasoning that the facts alleged to support those claims did not give rise to a violation of the law of nations. The court denied respondents' motion to dismiss with respect to the remaining claims, but certified its order for interlocutory appeal pursuant to §1292(b).

---

[26] Esther Kiobel, Individually and on Behalf of her late husband, Dr. Barinem Kiobel, et al. *v.* Royal Dutch Petroleum Co. et al. Certiorari to the United States Court of Appeals for the Second Circuit: No. 10–1491. Argued February 28, 2012—Reargued October 1, 2012—Decided April 17, 2013.

The Second Circuit dismissed the entire complaint, reasoning that the law of nations does not recognize corporate liability, 621 F. 3d 111 (2010). The United States Supreme Court granted certiorari and after listening to oral arguments affirmed the judgment of the Appeal Court.

The question therefore is whether to sue Chief Executive Officers (CEOs) and other responsible individuals within the corporation for their direct involvement in human rights abuses, since the courts have ruled that the law of nations does not recognize corporate liability.

The U.S. State Department of state was alleged to have at one point attempted to defeat a lawsuit alleging genocide and environmental damage that had been filed by Bougainville landowners against Rio Tinto's operation of the Panguna mine. State Department officials wrote to the judge hearing the case saying that airing of the class-action suit would affect U.S. relations with Papua New Guinea. The government of Papua New Guinea also tried to block the lawsuit, according to a report of November 30, 2001 by the *Australian Broadcasting Corporation*.

The issue of access to remedy for the victims of corporate abuse requires urgent attention. An obvious priority would therefore be to strengthen judicial institutions in the countries where abuses take place. However, it is also critical that the judiciary in multinational corporations' 'home countries, such as Canada, hear cases involving their companies in foreign countries, especially when the victims lack other viable options[27].

In view of the difficulties in bringing cases against corporations for human rights abuses and environmental pollution, an alternative development is and may likely be to inspire international regulation of corporate conducts in order to enforce good corporate behavior of extractive companies in the developing countries.

2. GENDER INEQUALITY IN NATURAL RESOURCES MANAGEMENT

The environmental despoliation and fatigue, bionetwork destabilization, and human rights abuses that consist of unlawful arrests and illegal incarcerations, forced-labor, forced-displacements or arbitrary relocation of residents, rapes and extra-judicial killings with manifest linkages to ethnic cleansing and genocides meted out to local residents of these natural resources development jurisdictions; the utilization of income from natural resources exploitations to fuel and sustain conflicts; mordant poverty, illegal arms trafficking by clandestine operatives

---

[27] [Refers to Cambior lawsuit re Guyana, Anvil Mining lawsuit re Dem. Rep. of Congo, Copper Mesa lawsuit re Ecuador, HudBay Minerals lawsuits re Guatemala. Also refers to Barrick Gold, TSX (part of TMX Group).

supplying arms to militaristic parties; acquisition and usage of weapons of mass destruction by warring groups and money laundering, untimely death of innocent citizens through environmental pollution and human rights abuses to mention a few affect both the male and female gender, however it is more exacerbated with the female gender the world over. This is so because when these things happen, it is the women who are left behind to pick up the pieces after the men have been taken away or killed.

The term gender refers to socially constructed roles and social expectations of individuals, based on their sex as male or female. Using the biological differences between men and women, the society establishes institutions, rules and regulations that determine what is expected of individuals, what their rights and privileges are, what resources they can access, whether or not they can participate in societal decision-making, and so on. In other words, the power of an individual is derived from the society in which he or she resides, temporarily or otherwise, and the way gender is constructed in that society. Like other categories of differentiation, such as ethnicity, race, class and language, gender determines a person's life chances, and shapes their ability to participate in the socio-economic and political life of a community or society.

The societal constructs about gender brings about such disparities as unequal opportunities in employment, unequal rewards for similar work, unequal access to various kinds of productive resources, and an unequal capacity to participate in and influence decisions that shape the development process.

The existing gender relations the world over favor men and stereotypes based on gender are deeply ingrained in society. Women's status is traditionally low, and they are mostly excluded from political and cultural life. Women's right to speak in public unlike the men is marginal, yet the women play a key role in the sustenance and development of humanity and the human society. This aside, in many developing countries hosting the extractive companies, education and cultural norms do not encourage women to speak out in public, yet they constitute around 53 per cent of the population; are responsible for children upbringing, unpaid domestic works such as cooking for the family, washing, fetching water and fire woods; and account for about 90 per cent of agricultural labor doing most of the farm work, transporting and processing of farm and forestry products. As farmers, they can observe the changes that are taking place in their environment, such as the impoverishment of the soil and activities associated with climate change phenomenon resulting from the activities of the extractive companies, in the fields where they work, and the environment where they live but because they are excluded from public debate, it is hard for them to help find solutions to these problems. Women's vulnerability is admittedly exacerbated by discrimination, above average unemployment, the absence of property rights and poverty. By prohibiting them from speaking out in their communities, women are thus denied the right to sustainable livelihoods.

Gender is cross-cutting issue in natural resources management and globally there are abundant evidence which suggest that while the benefits of extractive industry projects are captured primarily by men while the women often bear a disproportionate share of social, economic, and environmental negative consequences. Gender inequality and the disempowerment of women and girls have contributed significantly to the rampant discrimination against the females in the world today and this has transcended to the extractive sector where the host communities, multinational natural resources development companies and the host governments alike participate in the negation of women in decisions regarding the management of the extractive sector.

The male gender has better access to benefits of employment in the extractive sector, while the costs such as family/social disruption fall most heavily on women. It is worthy of note that if sustainability of extractive industries could be achieved there is need to identifying biases against the female gender and define effective strategies to ensure equal participation in decisions and sharing of the benefits.

The host communities, the host and home states and the extractive companies as a matter of utmost necessity must mainstream gender into their natural resources management policies and practices. There is a need to better understand and address how mining, oil and gas differently impact women and men. Achieving the development gains that extractive industries potentially holds out in particular for women depends on the understanding and management of such risks.

There is an urgent need therefore for the host and home states and the extractive companies to develop new prospects to ensure that women in extractive host communities increasingly become critical stakeholders who must participate in community consultations, discussions and decision-making. The female gender must be recognized as critical partners in sustainable extractive sector management in order to guarantee enhanced development opportunities, reduced poverty, and improved prosperity for the female gender.

For women in particular, extractive industries can provide opportunities for a better life, including increased employment opportunities, access to revenues, and expanded investment in the local community. Women-led businesses have flourished in other economic sectors and can also flourish in the extractives supply chain. Working with and investing in women by the extractive sector also makes good business sense since many companies are recruiting women to drive trucks and operate machinery, having discovered that women employees have patience leading to more impressive safety records and reduced maintenance of equipment than their male counterpart.

Understanding and acting on the gender dimensions of the extractive sector means including women in community-level project consultations, and national-level policy dialogues on extractive industries to ensure that men and women have equitable access to the benefits derivable from the development of their God-given natural wealth, and that neither is disproportionately placed at risks or discriminated against.

There is need for women to have equitable access to jobs, education, and participation. They must be included in making the decisions that affect their lives. Gender-sensitive consultation is essential to ensure that analysis; training and policies in the extractive industries not only meet the needs of women, but enhance their well-being.

The inability to create enabling environment and subsequent discrimination of women from effective participation in the extractive industries has resulted in the following subliminal and yet sordid consequences:

- Lack of voice and representation in the formal decision making process.
- Loss of ownership or use of fertile land or gardens.
- Loss of water resources and depleted fish stocks.
- Limited control over productive resources.
- Rise in violence and sexual abuse as a result of domestic disputes, alcoholism, drug use, or gambling.
- Rise in prostitution and HIV/AIDS and other STDs.
- Poor working conditions and incidences of sexual abuse for women in the project workforce.
- Environmental damage such as loss of forest and water sources and/or airborne or noise pollution which impacts women's lives and livelihoods.
- Loss of safety and security due to influx of construction workers.

The activities of the Bretton Woods institutions through SAP and liberalization of investment regimes have resulted in loss of jobs leading to increased poverty. The World Bank Group whose purpose of being is to alleviate poverty need to engage the host and home states and the companies to develop their extractive industries so that they become engines for economic growth and poverty reduction. It needs to promote equitable and inclusive sharing of benefits from the extractive sector. Having understood the negative social, economic and political consequences of the extractive companies activities, the Bretton Woods institutions must work with stakeholders to reduce potential environmental, social, and economic risks, raise awareness of the gender dimensions of the extractive industries, to ensure that all Bank-supported projects consider the needs and contributions of both men and women.

Providing women opportunity to participate in the extractive sector management will:

1. Enhance employment opportunity for the women in the extractive sector which in turn will benefit the community. This is so because where women have access to employment, or are empowered regarding household finances, evidence shows that they are more likely than men to invest in education, health, and nutrition for their families. Women have a better track record of starting successful business and repaying micro-credit loans, and show a greater willingness to respect safety and environmental safeguards. But where women have decreased access to employment, and to cash, families suffer more. Opening job opportunities to women therefore can increase productivity and reduce costs in the extractive sector as the women are often more likely to follow established rules, obey health and safety regulations, and can be more reliable employees.

2. Improve the opportunity for women to be involved in community consultations and decision-making in the extractive sector. Women involvement in community consultations to decide priorities for investment and spending of communal resources from the extractive sector will lead to more sustainable investment and spending and there is more likelihood of having sustainable development impacts.

3. Reduce community disruption and protest and can create a more predictable business environment with fewer production disruptions, thus avoiding cost increases and loss of income. Predictable environment is one of the driving forces that attract foreign direct investments to a country and such investments are necessary for job creation and economic growth as well as poverty alleviation.

3. INDUSTRY SKILLS GAP IN AFRICA

The US Geological Survey Annual Review of 2012 informs that with an estimated budget of US$3.4 billion, exports of mineral products and fuels accounting for up to 38 per cent of total exports; 12 per cent of world's crude oil reserves; well over a third of world's bauxite, gold, uranium and chromite; 88 per cent of world's diamonds; and 95 per cent of world's vanadium, Africa's extractive sector is a major economic sector that can generate longer-term social and economic benefits for the citizens.

According to the Deutsche Bank, US Geological Survey (USGS) and the World Nuclear Association, disaggregating these resources by regions provides the following resource profiles:

- North Africa- Phosphate 32 per cent of total world endowment.
- West Africa- Bauxite 40 per cent, Uranium 5 per cent and Iron ore 4 per cent of total world reserve.
- Central and Southern Africa- Platinum 88 per cent, Chromium 84 per cent, Diamonds 60 per cent, Cobalt 49 per cent, Gold 40 per cent, Uranium 13 per cent, and Copper 5 per cent of total world reserve.

With the above quantum of resource wealth, Africa has a great opportunity to broaden its economic growth potential and reverse its mordant poverty streak. It is a clearly known fact the world over that skills, and the institutions that affect the extractive sector level capabilities, constitute the most important determinant of economic benefits and the possession of specialized human capital adds value for local suppliers, creates a large number of direct and indirect jobs, and builds governance capacity.

According to the *Anglo American 2012 Sustainable Development Report*, economic value retained through employment and local suppliers accounts for 66 per cent of the total value created through minerals extraction. Unlike taxes, royalties, fees, and other revenues paid to government, skills development enables more value created from minerals extraction to be retained locally. Greater skills capacity enables higher levels of local employment and local procurement, in turn promoting inclusive growth and community empowerment. However, this possibility is encumbered by the lack of specialized industry technical expertise both in terms of numbers and quality, which has become a major access restricting gate, obstructing the potential for more well-paid jobs and local source of labor supply to extractive companies operating in Africa.

Take the example of De Beers that moved many of its downstream diamond activities from the United Kingdom to Botswana, making more diamonds available locally thus shifting more than US$6 billion worth of annual rough diamond sales from London to Gaborone creating an additional 3,200 manufacturing jobs in Botswana since 2007 and establishing 16 locally-based diamond buying companies.

To maximally benefit from the extractive sector, Africa's growing workforce should be able to capitalize on direct and indirect employment opportunities generated by extractive sector. However, the global extractives sector is characterized by very high investment risk and capital-intensive activities and requires high levels of and wide ranging skills in order to function effectively and to maximize the potential broader growth and development benefits. Lack of relevant skills therefore constrains local suppliers in upgrading firm-level operational competitiveness, meeting technical requirements, instituting innovation and adopting world class sustainable operations practices.

The skills gap analysis of the mining sector carried out in Zambia in 2012 concluded that the country was short of about 540 skilled workers. On the demand side, the surveyed companies employed 9,978 skilled workers, including 1,636 graduates, 1,427 technologists, 970 technicians and 5,943 crafts persons, out of a total of 32,515 unskilled and skilled workers in the sector. Some 300 workers were expatriates and mostly employed in technical and managerial positions. The study estimates that approximately 11,000 skilled workers will be needed by the companies surveyed in the coming five years, just to maintain the current level of production.

It was also observed that in Chile, the annual production of copper per worker was almost seven times greater than in Zambia but the lower level of production in Zambia could not be explained solely by variables like scale of operations, resources and equipment, however, the low level of productivity was attributed in large part to skills gaps that are rooted in weak technical and vocational training, in Africa.

For instance, despite increasing industry demand, graduates in engineering, manufacturing and other technical fields are scarce in Africa which parades the highest share of social science and humanities graduates in the world estimated at 70 per cent, compared to 53 per cent in Asia (OECD, 2012). To reverse the trend of skill gaps among African youths in the extractive sector requires expansion of the capacity of technical colleges, universities and secondary school systems to provide specialized programs of appropriate magnitude and quality.

At the tertiary level, the mostly generic-based extractives-related programs with little laboratory capacity must be strengthened. At secondary level, the culture of weak preparation in the sciences that prevents many students from undertaking engineering and science degrees at the tertiary level must be reversed. The introduction of quality vocational training for electricians, technicians, and heavy-machinery operators, to service the extractive sector skills demand would definitely add to close the skill gaps in the extractive sector in Africa.

4. INVESTMENT TREATIES

Transnational investment involves the transfer of tangible and intangible assets from one state to another in order to use that recipient state to generate wealth under the total or partial control of the owner of the assets. The state where the assets are transferred from is referred to as "home state," while the state where the assets are transferred to, is referred to as "host state". Formally, emphasis on transnational investment was on tangible property but now it is generally accepted that intangible property such as intellectual property rights, patents, copyrights, know-how, etc. fall within the protection of international law. A transaction such as a licensing agreement will therefore fall under foreign direct investment[28].

Transnational investment is not "a child of today" and can be traced to the pre-colonial era. However, following the Second World War, transnational investment became more prominent in the international scene[29]. This was given further impetus by the achievement of independence status by former colonies[30].

Transnational investment, during the colonial periods, did not generate as much controversy as it does now. This could have resulted from the fact that most of the investments prior to independence of the colonies were by investors from the colonists.

The colonial governments provided security and protection for these investments through the use of force (gunboat diplomacy). On attainment of independence, the newly independent states sort to assert their recently acquired sovereignty over natural resources[31], social, and economic life of their states.

Using their newly acquired sovereign power, the new independent states began nationalization assault on the assets of the foreign investors. In the years 1946-8, the new socialist countries (Poland, Czechoslovakia, Yugoslavia, Hungary, Bulgaria, and Romania) seized foreign properties and undertakings. Then, developing countries took the lead, in an impressive crescendo: Iran in 1951 (with the famous nationalization of the Anglo-Iranian Oil Company, mainly in the hands of British capital; the downfall of Prime Minister Moussadegh in 1954 and the concurrent agreement for compensation laid the British claim at rest); Egypt in 1956 (with the no less famous nationalization of the Suez Canal Company, the ramifications of which are all too known); Cuba in 1959; Sri Lanka in 1965; Tanzania in 1966; Bolivia in 1969; Algeria in 1971; Somalia between 1970 and 1972; Chile in 1972; Libya in 1978[32]. From 1960 to 1976 there were at least 1,369 reported instances of nationalization, affecting first property associated with the former colonial powers, particularly Britain and France, and then increasingly American owned property[33].

---

[28] See Sornarajah, The International Law on Foreign Investments, (1995) p.4.

[29] A. Cassese, International Law in a Divided World, (Clarendon Press, Oxford 1994) pp. 355-356.

[30] Ibid.

[31] UN Resolution 626 (VII) of 2 December 1952, was the first to recognize the rights of people to freely use and exploit their natural wealth and resources inherent in their sovereignty.

[32] , See Cassese pp 346. *Mutatis mutandis*

[33] UNCTC, *Transnational Corporations in World Development: A Re-Examination.* E/C.10/38 of 20 March 1978, 65

The concept of 'common heritage of mankind' as a guiding principle for the exploitation of natural resources which was first propounded in international forum in 1967, as a standard to establish a new regulation for the exploration and exploitation of the high seas[34] was brought to play. As at that time the notion of 'Permanent Sovereignty over Natural Resources" was very familiar and public debate was largely dominated by related ideas concerning a "New International Economic Order"[35].

The newly independent states saw the multinational corporations as agents of their home states and precursors to colonialism in the first place. They feared that if the activities of the multinational corporations are not curtailed, managed and controlled, that may result in total economic and political abrogation and eventual loss of sovereign powers by the newly independent sovereign states.

These newly independent states, mostly in the developing countries, therefore began to challenge the fairness of the existing international laws which they feel did not favor their course. On the other hand, the former colonists who helped fashion the laws oppose the demands of their former subjects. Due to the above struggle, one can safely conclude that the existing international investment law is not totally acceptable to the entire world community. One group cries foul to the existing regime and the other cries foul to calls for changes in the existing investment practices and treaties. Who will settle this matter and with what investment regime? One may be rightly tempted to look to international organizations for help.

From the above, it is evident that new transnational investment laws and regulatory frameworks that will meet the needs of both contending parties are needed. It is also evident that the host state, the home state, and international organizations will work together to facilitate the development of this new transnational investments rules. The problem with this arrangement is that the international organizations that are likely to influence decisions on this matter seem to always stand in favor of the articulated interests and positions of the capital exporting states at the expense of the capital importing states[36].

---

[34] Formally championed by the Maltese ambassador Arvid Pardo who noted in his speech of 1November, 1967 to the First Committee of the General Assembly, that the notion had been used previously, in July 1967, by the 'World Peace through Law Conference', an international gathering of private persons.

[35] Eva Paasivirta, *Participation of States in International Contracts,* (Finnish Lawyers' Publishing, Helsinki 1990) pg. 24

[36] World Bank, Legal Framework for the Treatment of Foreign Investment (Vols. 1and 2, 1992). Also, IMF, IRBD, OECD, GATT, MIGA, etc. are always in favor of capital exporting states.

Having provided a synopsis and a general background of problem of protection of transnational investors in the extractive companies, we now proceed to discuss first, the sovereign power of states under international law and how it enhances or militates against protection of foreign investments.

4.1 Sovereign Powers of States under International Law

Before any discussions on the sovereign powers of States under international law, it is important, first, to provide a definition of the term "Sovereign Powers of States". The Sovereign power of a State or Sovereign Prerogative of a State, as it is sometimes referred, is "that power of a state to which none other is superior or equal, and which includes all specific powers necessary to accomplish the legitimate ends and purposes of government[37].

Sovereign power or Sovereign Prerogative provides for State immunity, which developed from the personal immunity of sovereign Heads of States. The personal immunity of sovereign Heads of States is known as "Sovereign Immunity" and this doctrine could be traced back to the era before the twentieth century[38]. During that time individuals and corporations were not regarded as subjects of international law, only sovereign states were and hardly could any country in the world permit its courts to entertain actions brought by a private citizen against foreign sovereigns.

Under international law, all sovereigns were considered equal and independent[39]. The traditional view of immunity was set out by Chief Justice Marshall of the US Supreme Court in *Schooner Exchange v. Mcfaddon* (1812). This case resulted when some United States nationals began to claim ownership of the ship, *"Schooner Exchange"*, a French ship that was on a port-call in the US. Following the argument of the US Attorney General, Chief Justice Marshall stated that: *"the full and absolute territorial jurisdiction being alike the attribute of every sovereign and being incapable of conferring ...... to his independent sovereign station, though not expressly stipulated, are reserved by implication, and will be extended to him[40]."*

According to Lord Campbell, in the case of *De Haber v The Queen of Portugal* [1851], 17 Q.B. 171, 20 *"to cite a foreign potentate in a municipal court ....is contrary to the law of nations and an insult which I am entitled to resent"*.

---

[37] Henry Campbell Black, Black's Law Dictionary 6th ed. (1990) page 1396; *AETNA Casualty and Surety Company v. Bramwell,* D.C.; or 12 F. 2d 307, 309.

[38] See Philip Wood, Law and Practice of International Finance page 93.

[39] See Timothy Hillier, Public International Law (1994) P. 166.

[40] See Schooner Exchange v. Mcfaddon [1812] 7 Cranch 116.

Besides, some South American countries like Colombia and Venezuela have embedded in their constitutions, objections to foreign governing law, submission to foreign courts and waivers of immunity[41]. The adoption of these measures by these states could be traced to their adoption of 'Calvo doctrine' formulated by an Argentinean jurist, named Carlos Calvo, in 1868, and the Drago doctrine that followed in 1902.

The **Calvo Doctrine,** named after Carlos Calvo, an Argentine jurist, arose from Calvos's ideas, expressed in 1868 as necessary to prevent the abuse of the jurisdiction of weak nations by more powerful nations. The Doctrine is a foreign policy doctrine which holds that jurisdiction in international investment disputes lies with the country in which the investment is located.

The doctrine has since been incorporated as a part of several Latin American constitutions, as well as many other treaties, statutes, and contracts. The doctrine is used chiefly in concession contracts and the clause advocates giving local courts final jurisdiction and to obviate any appeal to diplomatic intervention.

The Calvo Doctrine also advocates the prohibition of diplomatic protection or armed intervention before local resources were exhausted, therefore, making an investor, to have no recourse but to use the local courts of the investment country, rather than those of their home country to seek redress in investment disputes. As a policy prescription, the Calvo Doctrine is an expression of legal nationalism which has been applied throughout Latin America and other areas of the world.

The **Drago Doctrine** which is a narrower application of Calvo's wider principle was announced in 1902 by the Argentine Minister of Foreign Affairs Luis María Drago. The Drago Doctrine itself was a response to the actions of Britain, Germany, and Italy, who in 1902 blockaded and bombarded sea ports in response to Venezuela's massive debt, acquired under President Cipriano Castro.

In addressing the Monroe Doctrine and the influence of European imperial powers, it set forth the policy that no foreign power, including the United States, could use force against any Latin America nation to collect debt. In 1904, the Roosevelt Corollary which asserted the right of the United States to intervene in Latin America in the interests of American business and Latin American independence from European powers was issued by the United States in response to the Drago Doctrine.

---

[41] Columbia does not permit submission to foreign law or forum. See Art. 127 of the 1961 Venezuelan Constitution.

However, a modified version of the Drago Doctrine by Horace Porter was adopted at The Hague in 1907, adding that arbitration and litigation should always be used first.

State immunity also prohibits the interference by one nation in the internal affairs of the other nations[42]. State immunity provided states the power to do absolutely anything. Immunity attached to all actions of sovereign states. This was then regarded as absolute sovereignty.

As set forth in the definition of sovereign power of the state, states have powers necessary to accomplish the legitimate ends and purpose of government. However, as states became more involved in commerce, and joining international organizations, it was apparent that if absolute immunity continues to be attached to the powers of sovereign states, these powers could be utilized by the states, to the detriment of individuals and business corporations involved in transnational ventures in their various states or who enter into contracts with them. The issue created lack of universality in the definition of purpose of government.

4.2 Multinational Efforts to Create Transnational Investment Rules

State immunity provides that states have the power to do absolutely anything including expropriation of foreign investors' assets without anyone to question their activities. An argument was advanced that if states continue to expropriate foreign assets with impunity and without any form of checks; there will not be any protection for foreign investors' assets. This triggered the fear of frequent expropriation foreign investors' assets and this prompted quest for delimiting the absoluteness of state sovereign immunity, when trade and commerce are involved. This culminated in establishing a distinction between public and governmental acts of a sovereign (*acts jure imperii*) and its private or commercial acts (*acts jure jestionis*).

Along this line, the International Law Commission (ILC) Draft Articles favored restriction on state immunity. The following stipulations are contained in the ILC Draft Articles.

Draft Article 10 provides that a state will not be immune in respect of a commercial transaction. Draft Article 11 establishes that a state will not be immune from such matters as employment contracts and Draft Article 12 prohibits immunity in the case of personal injury or damage to tangible property. Therefore under the foreign immunity Act, a foreign state can be sued in the jurisdiction of another state where the nature of the suit emanates from commercial engagement.

---

[42] See Buck v. Attorney General [1965] Ch. 745.

Nationalization of foreign extractive sector assets cannot be avoided but can be managed to reduce the negative consequences associated with it. There are provisions under international law that attempt to mitigate expropriation risks of transnational investments. These international instruments have proven to be both inadequate and ineffective in dealing with problems associated with transnational investments in natural resources. As mentioned above, there is ideological difference between home states and the host states on the international law with regards to the protection and treatment of transnational investments. Some attempts have been made at drafting comprehensive code[43] but resulted in failure due to the ideological difference and other rifts prevalent in this branch of international law.

The United Nations set up a Commission on Transnational Corporations (UNCTC) to draft a code of conduct for the multinational corporations. This attempt has not been fully successful as there continue to be disagreement between the capital exporting and capital importing states. But some provisions of the UNCTC draft code have been adopted.

Article 7 of the code calls on the transnational corporations to respect the sovereignty and obey the laws of the host states, and the host states to honor in good faith, its international obligations. There is also a provision that calls for contract renegotiations when there is a fundamental change in circumstances under which the contract was negotiated. Besides, transnational corporations are required not to interfere personally or influence their home states to interfere in the domestic political affairs of the host states. Such interference is inconsistent with the United Nations Charter and Declaration of Friendly relations between states[44].

In addition to the above, transnational corporations on one hand are required to protect the environment, labor relations and prevent restrictive business practices as well as provide information to the public on financial and other matters relating to the operation of the corporation.

The host state on the other hand are required to obey international legal rules relevant to the treatment of transnational corporations; to compensate for nationalization of assets; provisions of jurisdiction; and dispute settlement.

---

[43] See Sornarajah, p. 187.

[44] U.N.C.T.C Articles 16-20. A precursor to this provision was the fear of the developing nations that multinational corporations will use their economic power to influence their home states to intervene in the internal politics of the host state. An example was the interference in Chile which resulted in the Overthrow of President Allende. Another was the interference in Nicaragua by the US. See [1984] *ICJ* Reports, 352

There are other international instruments espoused by the World Bank to protect transnational investment in the extractive sector. The first is the establishment of International Center for the Settlement of Investment Disputes (ICSID)[45].

The other World Bank's instrument to protect transnational investment is the establishment of Multilateral Investment Guarantee Agency (MIGA). This agency was established in 1985 and has been in operation since 1988. MIGA serves nationals of member states or legal persons who are incorporated and have a seat in a member state or legal persons with most of their assets in a member state.

The main aim of MIGA is to provide protection for direct foreign investments. The scope of coverage includes currency transfer; expropriation; breach of contract; war and civil disturbances; and the Board of Directors may extend coverage to other non-commercial risks[46]. These protective measures have proven to be ineffective in protecting foreign investments. As a result, capital-exporting states have resorted to bilateral investment treaties (BITs).

4.3 Bilateral Investment Treaties (BITs)

Bilateral Investment Treaty (BIT) in transnational investment is an investment agreement between two states to protect the investment of capital exporting states by capital importing states. It is an agreement between two states on the promotion and protection of foreign investments by the host state. The first bilateral investment treaty in the world was signed on November 25, 1959 between Germany and Pakistan.

BITs sprang up as investment protection measure because multilateral agreements were not adequate and effective enough to protect the investments against expropriation risks. Both developed and developing states were always at disagreement on the best way to direct the course of transnational investment laws. For this reasons, states had to resort to the second best solution by entering into BITs to ensure the existence of definite rules relating to transnational investments[47].

After the Second World War, transnational investment became very prominent in international arena. The achievement of independence by former colonies heightened the tension in transnational investments. The

---

[45] ICSID founded on the convention of 18 March 1965, adopted by over 140 states by December 1997.
[46] MIGA Convention, Art 11(b) and (c)
[47] See Sornarajah p. 234.

newly independent states began to challenge the provisions of the existing international law, especially in area of protection of foreign investments by the host states and nationalizing foreign assets of former colonists.

BITs are said to be one-sided agreements in that the powers of the contracting states are not at equilibrium. There were no real agreeable effective and adequate instruments developed to protect foreign investors' assets. There was constant disagreement between capital exporting developed countries and capital importing developing states as to the best way to protect these investments. The capital exporting state, usually more powerful than the capital importing state, enters into this treaty to protect the investments covered in the agreements with the host state.

BITs regulate investments either when both states import and export capitals or when one exports and the other imports capital. This led to numerous but unsuccessful efforts of the world community to develop multilateral rules in area of transnational investments. Some of these efforts met with continued opposition from both developed and developing states.

Due to the attendant lack of successes in developing internationally agreeable multilateral investment regime, BITs became a new phenomenon as a way to establish at least a rule for the protection of foreign investments against expropriation by capital importing host state.

Some proponents have suggested that BITs would lead to the creation of complete customary international regime in the field of foreign investment protection. Others argue to the contrary. In order to determine whether or not BITs will eventually create a complete customary international law in transnational investment protection, this writer reviewed and analyzed cases arising from BITs, literally work of experts, and the provisions of the U.N, and other international organizations.

Findings are that many states are signing BITs as a means of promoting and protecting foreign investments. Because of this, one could agree that BITs would become a complete customary international regime in the field of foreign investment protection.

BITs are offer states a deliberate method by which to create international law and are fast becoming the most important source of international law[48]. Therefore most of the law binding a state will take the form of bilateral treaties between two states or multilateral treaties[49].

---

[48] Martin Dixon & Robert McCorquodale, Cases and Materials in International Law, (2nd ed. 1991) p.24.
[49] Ibid.

On the other hand, one could conveniently argue that BITs do not muster the homogeneity required to assume the status of customary international regime. This position is supported by the fact that for BITs to become a complete customary international regime in the field of foreign investment protection there must be transparency and *opinion juris* in state practices.

BIT aims at promoting investment between two nations and transfer of technology to the host states. Secondly, it aims at providing protection against expropriation of foreign investments. This leads to internationalization of the treaty by subjecting the parties to arbitration before foreign tribunals, and choice of law, as well as stabilization clause.

[50]There is a claim that BITs boost investors' confidence in the host state and as a result more investments take place. According to Sornarajah, the principal reason for the developing countries concluding such treaties is the belief that the treaties will lead to greater investor confidence by dispelling any previous impressions of risks associated with the host state[51]. However, evidence suggests otherwise[52]. At the UNCTAD inter-governmental expert group meeting of 28-30 May 1997, a representative from Nigeria said that "although many African countries sign investment treaties they do not get foreign investments. Investment flows are determined by access to markets, social and political conditions".

Also Ms Li Ling (Deputy Director-General, Treaty and Law Department, Ministry of Foreign Trade and Economic Cooperation) from China observed that there is no evidence that the investor will deem the host country unsafe if that country did not conclude a BIT with their home state.

Besides, she noted that US is the biggest investor in China behind Hong Kong, Macao and Taiwan but SINO-US BIT has not been concluded yet. Further, the Brazilian expert representative at the meeting noted that BIT was not important as an investment incentive and that market size was more important.

In the same line of argument, the Kenyan expert noted that Africa had attracted little FDI despite liberalization, and Nigeria attracted 60% of Africa's FDI because of its large market and oil resources. The Indonesian expert similarly noted that three of the largest five countries from where foreign investments came to Indonesia did not have BITs with Indonesia. That Bit was only a tool to assure the home state of investor's security. Further, the

---

[50] See Sornarajah p. 232.
[51] See Sornarajah p. 236.
[52] See the UNCTAD inter-governmental expert group meeting (28-30) May 1997. See the comments by Professors Jean-Luc Le Bideau and M. Sornarajah

delegate from Japan agreed that BITs have a "very limited role" in promoting FDI to developing countries. He mentioned that Japan had concluded BITs with only four countries and that the major recipients of Japanese FDI were Southeast Asia region and China and Japan had no BIT with Southeast Asian countries, only with China. He concluded by saying that "a stable political and social condition was the most important investment factors". In affirmation, the US delegate said that "the US entered into BITs to protect US interests abroad and have its investors treated fairly. For the US, at the heart of BITs are three elements: non-discrimination to FDI, protection of investor's rights, and enforcement mechanisms for the first two"[53].

In furtherance of discussion on BITs, it is necessary to look at the question as to whether BITs can become complete international regime to protect foreign investment. In this regard therefore, it is of fundamental importance to mention briefly the source of international law. This will help in the analysis to determine whether BITs can metamorphose into international law for the protection of FDIs.

International Law is created mainly in two ways: by treaty and by customs[54]. There are other sources of international law; however they may not be discussed in this book.

Treaties are agreements between two or more states, and are binding on the states involved if they have given their consent to be so bound[55]. Customary law is established by showing that states have adopted broadly consistent practices towards a particular matter and that they have acted this way out of sense of legal obligations[56].

The doctrine of *pacta sunt servanda* is the fundamental principle of the law of treaties[57]. This is established in Article 26 of the 1969 Vienna Convention on the Law of Treaties. It states *inter alia* that *"every treaty in force is binding upon the parties to it and must be performed by them in good faith."* Article 34 provides that a treaty does not create either obligations or rights for a third state without its consent.

BITs are treaties between two states and are not binding on a third state. Article 38 provides that nothing in Articles 24-37 precludes a rule set forth in a treaty from binding on a third state as a customary rule of international law, recognized as such. BIT is not a customary international law and cannot be binding on a third state. Article

---

[53] Ibid.
[54] A. Bradney et. al, How to Study Law (2nd. Ed. Sweet & Maxwell 1995) p.17; Supra note 11 page 816
[55] Ibid.
[56] Ibid.
[57] Vienna Convention on the Law of Treaties (1969).

60 provides that *"a material breach of a bilateral treaty by one of the parties entitles the other to invoke the breach as a ground for terminating the treaty or suspending its operation in whole or in part."*

When multilateral treaties are acceded to and ratified by states, they become legally binding on international community despite the fact that not all states must have ratified the treaty. The only way a state may claim reservation to the binding legal obligation is if the state has continuously objected to the treaty from the formulation stage. This was evidenced in the Nicaragua Case (Meritis)[58].

Article 38(1) (b) of the statute of the I. C. J refers to international custom *"as evidence of a general practice accepted as law*[59]." International Customary Law is created when two conditions are fulfilled:

(a) a practice has to be followed by more than two states, and

(b) This practice has to be accepted by law (opinio juris)[60].

BIT is gaining a wide recognition in the field of FDI protection as many states are rushing to sign and ratify the BIT. Because of the acceptance of BITs by many states as a means of promoting and protecting foreign investments, it could be argued that BITs may become complete customary international regime in the field of foreign investment protection. But first, BITs must have to undergo transformation into multilateral treaty, which must have been ratified by U.N Member states, and there have been a consistent practice by more than two states (unlike in BITs) and accepted as *opinio juris*[61].

In the Nicaragua Case, the court opined that *"where two states agree to incorporate a particular rule in a treaty, their agreement suffices to make that rule binding upon them; but in the field of customary international law, the shared view of the parties as to the content of what they regard as the rule is not enough"*[62].

---

[58] See *Nicaragua v United States, I.C.J.* Reports 1986, p.14.
[59] See Charter of the United Nations Chapter 2 Art. 28.
[60] Hans Van Houtte, *"The Law of International Trade"*, (Sweet & Maxwell 1995) p. 9, para. 1.10
[61] D.J Harris, Cases and Materials on International Law, (5th ed. Sweet and Maxwell, London1998) p.45.
[62] Supra note 34.

In Texaco v. Libya[63] arbitrator Dupuy ruled that it is generally accepted that international law allows states to nationalize foreign assets. Nationalization by states of foreign assets is therefore an accepted customary law by the world community.

Bilateral investment treaties have come under the jurisdiction of domestic courts, despite the efforts to internationalize them through provision for arbitration clause under foreign law and through international arbitration centers.

The WB ICSID arbitration, the most frequent arbitral system used by investors under BITs, was designed to be insulated from interference by domestic courts, because of the belief held in some quarters that the domestic laws of developing countries are not matured or developed enough to deal with investment issues or because of bias against foreign investors, in addition to the level of corruption in the judiciary. The ICSID Convention precludes appeals to domestic courts in the seat of the arbitration and provides for an internal ICSID Annulment Committee to review the arbitral awards. Despite this fact, domestic courts still play very important roles with respect to BITs and such roles include:

1. The review of the constitutionality of BITs.
2. Redress breaches of BITs.
3. Interpretation and application of BITs provisions.

The domestic courts have examined claims relating to challenge and disqualification of arbitrators set up under investment treaties. An example of such case is the *Republic of Ghana v Telekom Malaysia Berhard*[64], brought within the ambit of Ghana-Malaysia BIT. In this case, the District Court of The Hague where the arbitrators were seating to arbitrate found that there was a functional conflict of interest between the position of an arbitrator in the Ghana case and the position of an advocate seeking the annulment of an award in another BIT case before ICSID. Another illustrative case is the *Poland v Eureko*[65], in which the Court of First Instance in Brussels rejected a request to disqualify an arbitrator from an arbitral tribunal constituted in Brussels within the framework of

---

[63] Texaco v. Libya (1977) 53 *I.L.R.* 389; (1978) 17 *I.LM.* 1

[64] *Republic of Ghana v Telekom Malaysia Berhard*, District Court of The Hague, 18 October 2004, Challenge No. 13/2004; Petition No. HA/RK 2004.667; and Challenge 17/2004, Petition No. HA/RK/2004/778, November 5, 2004, available at *TDM*, Vol. 2 - issue 1 January 2005

[65] The Judgment of Court of First Instance of Brussels on challenge to arbitrator, of 22 December 2006. available at http://ita.law.uvic.ca/

the Netherlands-Poland BIT. The judgement of the Court of First Instance was upheld by the Brussels Court of Appeals[66].

Domestic Courts also in some jurisdictions, review the constitutionality of BITs or laws approving them. This review power may be entrusted to a special constitutional court or to the ordinary courts. Decisions on the constitutionality of investment treaties have been rendered in Canada and Colombia.

In Canada, there have been at least two constitutional challenges of the North Atlantic Free Trade Agreement (NAFTA) investment chapter. On March 2001, the Council of Canadians, the Canadian Union of Postal Workers and the Charter Committee on Poverty Issues filed a notice of application in the Ontario Superior Court of Justice alleging that the dispute settlement provisions of Chapter 11 of NAFTA violated the Canadian Charter of Rights and Freedoms as well as the Constitution Act. The Claimants asserted that Chapter 11 of NAFTA deprives Canadian courts of the authority to adjudicate claims against the State by private parties, matters reserved to them by the Constitution.

They also argued that this chapter infringes and denies the rights and freedoms guaranteed by the Charter of Rights and Freedoms and the Canadian Bill of Rights, including those concerning fundamental justice, fairness and equality[67]. Justice Pepall of the Ontario Superior Court dismissed the application, acknowledging that though NAFTA arbitration 'lacks predictability', 'lacks total transparency' and there is no consistent mechanism for review of the decisions rendered by NAFTA tribunals'. Nevertheless, she dismissed these concerns, pointing out that 'a treaty is a bargain', and that her task is 'not to remedy unpopular provisions', but to determine whether Chapter 11 is in violation of Canada's Constitution. The Court of Appeal of Ontario confirmed this decision[68]. In addition to this first constitutional challenge, on May 2001, Democracy Watch and the Canadian Union of Public Employees filed a notice of application in the Ontario Superior Court of Justice seeking to declare NAFTA Chapter 11 unconstitutional and inconsistent with the Canadian Charter of Rights and Freedoms.

---

[66] Brussels Court of Appeal, Decision of October 29, 2007, R.G. 2007/AR/70, available at http://www.master-arbitrage.uvsq.fr/revue/vol2-2008/ ChroniqueLPA200803.pdf

[67] Council of Canadians, CUPW and the Charter Committee on Poverty Issues *v* the Attorney General of Canada, Ontario Superior Court of Appeal, Reasons for Judgment, July 8, 2005 available at http://www.international.gc.ca/trade-agreements-accords-commerciaux/disp-diff/cupw.aspx?lang=en.

[68] Court of Appeal for Ontario, Reasons for Judgment, November 30, 2006, available at http://www.international.gc.ca/trade-agreements-accords-commerciaux/disp-diff cupw.aspx?lang=en.

In Colombia, the Constitutional Court must first review all treaties, including BITs, approved by the Congress before the President appends his signature. During the constitutional review, citizens and public authorities have the right to intervene. Through this review, Colombia Constitutional Court has adjudicated on the constitutionality of Colombian BITs especially the constitutionality of the expropriation clause contained in many Colombian BITs.

Subsequently, the Court has found this provision inconsistent and in violation of Article 58 of the Constitution that authorizes the Colombian Parliament, by vote of an absolute majority, to decide not to pay such compensation for reasons of equity. Secondly, Constitutional court has found that the expropriation clause is inconsistent with Article 12 of the Constitution that requires an equality of treatment between Colombians and Foreigners since the absolute right to compensation benefits only foreign investors. The BIT obligates investors to turn to domestic courts before seeking arbitration in respect to administrative acts, as required by Colombian law, it is more appropriate to submit investment disputes to an investor-state arbitration mechanism.

Some BITs give domestic courts the initial opportunity to remedy any injustice committed against the foreign investor before a party can request international arbitration. In many BITs however, investors have an option to bring a claim to a domestic court or to an arbitral tribunal and domestic courts have the jurisdiction to review the State's fulfilment of its obligations under investment agreements.

When facing a tax claim before the French Cour de Cassation, the highest court in the French judiciary, the French Tax Administration adopted this position. The *Directeur general des impôts* asserted that the Panama-France BIT of November 5, 1982, which provides in its Article 8 for an investor-State arbitration under UNCITRAL rules, did not contain a substantial provisions directly applicable to nationals and companies and does not entitle them to a direct right of action before the courts'. Unfortunately, the Cour de cassation did not examine this issue. However, the *Conseil d'Etat*, the French higher administrative jurisdiction, in a decision, held that Article 3 of the Algeria-France BIT of 1993 relating to fair and equitable treatment and prohibition of unreasonable and discriminatory measures has only an inter-state effect and cannot cover a private person contesting the rejection of his visa request[69].

Some Philippine NGOs lodged a petition before the Philippine Supreme Court arguing that the investment chapter of the Japan-Philippines Economic Partnership Agreement violates the Philippine Constitution. Mainly, they asserted that this chapter violates constitutional limits on foreign ownership in some sectors like real estate, mass media, advertising, and public utilities.

---

[69] Conseil d'État, N°280264, Decision of December 21, 2007, available at http://www.legifrance.gouvfr/.

Similarly, the Supreme Court of Pakistan, in the *SGS* case[70] held that neither the Swiss-Pakistan BIT of 1994 nor the ICSID Convention had been incorporated into Pakistani law by legislation and, therefore, these two agreements could not be relied upon to confer rights on individuals.

In *Kessl v Minister of Lands and Resettlement et al*, the investors challenged the expropriation of their land by the Namibian government and asserted that the way in which the expropriation was carried out breached the Namibian Constitution, the land reform law and constituted discrimination prohibited by the Germany-Namibia BIT, given that it focused specifically on farms belonging to foreign nationals. In its decision of March 6, 2008, the High Court of Namibia referring to the Germany-Namibia- BIT to settle an expropriation dispute between three German nationals and Namibia, ruled in favor of the claimants.

Although the Court's decision was based on national law, it did rule that the BIT must be respected. It pointed out that: 'As German citizens, the three applicants are entitled to the same treatment as Namibian citizens in terms of the Encouragement and Reciprocal Protection of Investments Treaty which was entered into by the Republic of Namibia and the Government of the Federal Republic of Germany'. In the same vein, some decisions of the Mexican Supreme Court have referred to the non-discrimination rule provided in the NAFTA to confirm a solution handed down based on the Constitution.

Furthermore, BITs make reference to foreign investors' ability to enforce BIT rights through the domestic courts. This reference is an important element in showing that the treaty creates a private right of action. On 26 August 2008, the US Court of Appeals for the DC Circuit handed down an important decision relating to the applicability of an expropriation provision in the Treaty of Amity, Economic Relations and Consular Rights between the US and Iran in domestic legal proceedings. In this case, a US company claimed that Iran had expropriated its property without paying compensation, which is incompatible with article IV of the treaty. The Court maintained that it cannot create a cause of action before the US courts for a US company. The Court stated that according to US case law, there was a presumption that 'international agreements, even those directly benefiting private persons, generally do not create private rights or provide for a private cause of action in domestic courts'. Pointing out that the Treaty does not refer to the dispute resolution methods that would enable nationals from the two countries to ensure that conventional obligations are fulfilled; the Court of Appeal held that violations of these treaties must be addressed by direct negotiation between the States that signed them.

---

[70] Supreme Court of Pakistan (Appellate Jurisdiction), Société Générale de Surveillance S.A. *v* Pakistan, July 3, 2002, § 18-28, International Law Reports, Volume 129, p. 323. The Court also decided that SGS didn't make a protected investment within the meaning of the BIT.

Even when foreign investors decide to go for international arbitration, domestic courts still play a significant role. For instance, the domestic courts have been asked to prevent the investor from commencing or continuing an arbitral proceeding under an investment treaty usually through the *anti-suit injunction*, which is an order issued by a domestic court to the investor or the arbitral tribunal to stop the proceedings. In *SGS v Pakistan*, the Supreme Court of Pakistan issued an anti-suit injunction against SGS restraining it from taking any action to pursue its BIT arbitration[71].

There are many decisions illustrating domestic court intervention to grant provisional measures in relation to an investment arbitration proceeding. Domestic courts may assume this function even when the investor has decided to submit the claim to ICSID arbitration. ICSID rules allow parties to seek provisional measures from national courts if authorized by the applicable investment treaty[72]. The English Court held that it had no jurisdiction to grant an injunction in support of ICSID arbitrations in the absence of the disputing parties agreement and also that the Respondents were, in any event, protected in these circumstances by sovereign immunity legislation.

Non-ICSID investment awards can be challenged before domestic courts, usually the courts of the place of arbitration. Requests to set aside an investment award are made in such countries as Canada, the US, Belgium, and England, among others. Only in one case did a domestic court decide to annul an arbitral award rendered under an investment treaty. All these proceedings usually involved the application of local arbitration law on vacating arbitral awards.

However, some courts asked to hear a challenge have interpreted some investment treaty provisions. Some decisions deal with the interpretation of the investor-state clause. In *Czech Republic v CME B.V*, the Svea Court of Appeal interpreted the choice of law clause found in Article 8.6 of the BIT between the Czech Republic and Netherlands[73]. In *Czech Republic v European Media Ventures SA*, the English High Court clarified the proper approach to interpreting BIT provisions and determining the scope of the arbitral tribunal's jurisdiction under Article 8 of Czech-Belgo Luxembourg BIT. The High Court decided that this provision did not limit the arbitrators' jurisdiction to the issue of quantification but conferred substantive jurisdiction to determine whether compensation should be awarded.

---

[71] The ICSID Tribunal however in its Decision on Jurisdiction rendered on 6th August 2003, in ICSID Case No. ARB/01/13, disregarded the decision of the Pakistani Supreme Court.

[72] See Article 47 of ICSID Convention, and Article 39 (6) of the ICSID Arbitration.

[73] *Czech Republic v CME B.V*, Review by Svea Court of Appeal, 15 May 2003, 42 *ILM* 919 (2003), 90.

In the *Republic of Ecuador v Occidental Exploration & Production Co*[74], the High Court of Justice rejected Ecuador request to set aside an arbitral award rendered in London pursuant to the UNCITRAL Rules. Ecuador contended that the arbitrators had exceeded their jurisdiction by making an award on matters of taxation which, it claimed, were excluded by Article X of the BIT. However, after a careful interpretation of this provision, the court found that, although Article X of the Treaty excluded matters of taxation from the scope of the Treaty, an exception was provided in the BIT.

The use of international arbitration mechanism to resolve investor-State disputes in the extractive sector has increased exponentially over the past few decades. However, the experience of practitioners, arbitrators, academics, public officials, and policy-makers and advocates on the advantages, disadvantages, and viability of the current dispute resolution mechanism in Foreign Direct Investment reveals that there is complexity in international investment agreements (IIAs). These complexities include inconsistencies in the investment treaties themselves, inconsistencies in the interpretation of investment treaties; the independence and impartiality of arbitrators and the costs and time of arbitral proceedings; contradictions in arbitral awards; difficulties in reversing "erroneous" arbitral decisions, and equity in reciprocal enforcement of foreign arbitral awards; all of which lead to broader question of the legitimacy and adequacy of the system of dispute resolution with the current international investment treaties such as the BIT. Based on the above, the current systems have not mustered the capacity to successfully resolve investment disputes. There are perceived limitations of those mechanisms, and UNCTAD recommended five[75] broad reform alternative paths model framework in place of the current system. According to UNCTAD the alternative system should aim at:

- Promoting alternative dispute resolution;
- Tailoring the existing system through individual IIAs;
- Limiting investor access to ISDS;
- Introducing an appeals facility; and
- Creating a standing international investment court.

---

[74] *The Republic of Ecuador v Occidental Exploration & Production Co.*, High Court of Justice-Queen's Bench Division (Commercial Court), March 2, 2006.

[75] D. Gaukrodger and K. Gordon, Investor-State-Dispute Settlement: A Scoping Paper for the Investment Policy Community, *OECD Working Papers on International Investment, N°2012/3; K.P. Sauvant and F. Ort*ino, Improving the International Investment Law and Policy Regime: Options for the Future, Seminar on Improving the International Investment Regime, Helsinki, 10–11 April 2013 (hosted by the Ministry of Foreign Affairs of Finland).

The above UNCTAD recommended alternative systems are not enough to make any positive imputes, instead they are just another instrument of confusion. Instead, the role of domestic courts in the settlement of disputes involving foreign investors and host States should be re-examined as this could hold more potency in view of the increased level of activities in current framework.

BITs differ from one another in contents and provisions, and depending on the relationship that existed or exists between the two state parties and to a large extent on the nature of the business involved. Because of the lack of homogeneity in BITs, it is difficult to argue confidently that BIT will create a complete regime in the field of foreign investment protection.

For BIT to form a complete customary international law regime in the field of foreign investment there must be transparency in state practices and also *opinion juris*[76]. In his argument as to why BITs cannot muster a qualifying standard of comity to become a customary international law, M. Sornarajah postulates that "*the existence of incompatible objectives between the parties to the treaties, the failure of important states to engage in such treaty practice, and the variety of formulations that are included in these treaties make them unsuitable foundations for the making of customary international law*[77]". Also, Friedmann[78] noted the importance of customs in international law, but basically supported the view espoused by M. Sornarajah.

5. FINANCIAL MISMANAGEMENT AND CORRUPTION

Integrity, transparency and accountability are the tripod of good governance and the absence of good governance breeds corruption which forms the pivot of unsustainable socioeconomic development, poverty, misery, security challenges among others, with their attendant consequences.

Corruption is one of the greatest atrocities committed by the leadership of developing countries, but those enveloped by its callous poisonous effects have been drained of their guts, stamina, health and all vital signs to fight it, thereby expanding and perpetuating the horizon of despondence and agony among the citizens. Corruption has become the most talked about issue in the world, especially in developing countries where extractive industries operate, ranking ahead of poverty, education and health care.

---

[76] Ibid.

[77] Sornarajah, p.236.

[78] Wolfgang Friedmann, the Changing Structure of International Law, (1964) pp. 121-123.

The United Nations Development Program (UNDP) defines corruption as "the misuse of public power, office or authority for private benefit – through bribery, extortion, influence peddling, nepotism, fraud, speed money or embezzlement".

According to the former Secretary-General of the United Nations, Kofi Anan, "Corruption is an insidious plague that has a wide range of corrosive effects on societies; it undermines democracy and the rule of law, leads to violations of human rights, distorts markets, erodes the quality of life, and allows organized crime, terrorism and other threats to human security to flourish. Corruption hurts the poor disproportionately by diverting funds intended for development, undermining a government's ability to provide basic services, feeding inequality and injustice, and discouraging foreign investment and aid".

Financial mismanagement and corruption play critical role in the unsustainable management of the extractive sector, leading to agony of host communities and countries where the extractive companies operate.

One of the major causes of corruption is "wide authority imbued with minimal accountability". When one is given an excessive authority without corresponding measure to check or accountability, there is the temptation of becoming overbearing. There is a saying that "power corrupts and absolute power corrupts absolutely".

Another cause of corruption is that it takes long time to build integrity and establish credibility. Further, the administrative systems in some developing countries hosting the extractive industries have inbuilt weaknesses where government agencies that generate primary sources of revenues for the country are not effectively monitored.

There is excessive individual discretion and a system lacking structures to effectively monitor the exercise of discretion and hold decision-makers accountable; a system in which individual offices, departments or agencies operate in isolation from one another as against a system where individual elements operate in a coordinated fashion and communicate regularly with one another and tend to carry out mutual "monitoring" both of activities and individuals are apt to be corrupt.

Corruption is endemic in the extractive host countries because people want to take unfair advantage over others—to pay lower taxes, to get an appointment or promotion, to win a contract, or to get something done quickly, to avoid paying a fine or penalty.

Those politicians and officials who fear loss of office seek corrupt benefits as "insurance", especially when they can expect no pension; officials need extra money to maintain their standards of living if salaries have not been raised to match inflation, to meet commitments for housing, car, school fees, etc.; employees feel resentment over bad management or pay levels they think unfair. This may make them feel justified in making false expense claims or taking bribes.

The employees who refuse to participate in a corruption "racket" may be suspected and put under threat from their colleagues or superiors; some seek status, not only for having more riches than their colleagues but because corrupt officials may be admired by friends and family for their skills in outwitting authority.

Job insecurity is another factor causing corruption. This is a situation when an employee is not sure of having a job tomorrow to maintain the family.

Corruption occurs when organizations or individuals profit improperly through giving and receiving bribes, extortion, favoritism and nepotism, as well as embezzlement, fraud, conflicts of interest and unlawful monetary contributions to political parties as well as procurement and provision of substandard products and services, among others.

Through corrupt practices political leaders lose credibility and must buy support; legislation is shaped by the highest bidder, not by social needs; public accountability, law enforcement and bureaucratic functions are compromised and ineffective, political parties become patronage operations serving leaders' personal interests, judicial decisions are sold or politically manipulated, the press and civil society remain too weak to check the powerful, there is the depletion of national wealth. Corruption hinders the development of fair market structures and distorts competition, generates economic distortions in the public sector by diverting public investment into capital projects where bribes and kickbacks are more plentiful, inflates the cost of business, it also distorts the playing field, shielding firms with connections from competition and thereby sustaining inefficient firms and undermines people's trust in the political system, in its institutions and its leadership and it frustrates and produces general apathy among disillusioned consumers and weakens the civil society which in turn clears the way for despotism, demanding and paying bribes and those unwilling to comply often emigrate, leaving the country drained of its most able and most honest citizens

Corruption creates room for environmental degradation, lowers compliance with construction, environmental, or other regulations, reduces the quality of government services and infrastructure; provide access to careless exploitation of natural resources and ravaged natural environments.

Corruption leads to election rigging and bad governance; to lack of implementation of laws and policies, lack of rule of law, weak law enforcement institutions, lack of public confidence in government institutions, lack of transparency and accountability in government processes, weak electoral systems, breeds conflicts over state resources, denies the poor access to basic services, deprives the poor the opportunity to participate in governance, public distrust in governance, nepotism, wastes of social capital, inefficiency in public service, high cost of doing business, reduced economic growth, higher consumer prices, reduced competition, capital flight, diversion of state resources and revenue, bad procurement, abuse of official guidelines and abuse of privatization rules and the overall impacts on unsustainable products and services consumers, providing enabling environment for increased products and services prices but not quality and performance, and this has "contributed substantially to lowering the quality of life and well-being of the consumers who have become more and more impoverished.

Corruption creates an environment where citizens have become victims of unwholesome business practices resulting in avoidable deaths that have created orphans, widows and widowers, and single parenthood; maiming that have created disabilities and also loss of properties without compensation. All these culminate to expand the dimension of poverty, insecurity of life and property, the abuse of human rights and diminished happiness of the citizens.

Corruption leads to falling standards of education; the rich to make frequent overseas trips for medical care, countless of citizens losing their lives every day as a result of afflictions suffered from curable diseases; environmental despoliation, destruction and neglect or mismanagement of government infrastructure, kidnapping and militant activities among others.

Corruption reduces the overall wealth of a country since it can discourage businesses from operating and average income is about three times lower than in less corrupt countries; it also reduces the amount of money the government has to pay good workers and purchase supplies, such as books and medicine. It distorts the way the government uses its money, too. The result is that schools, health care facilities, roads, sewer systems, security forces, and many other services that governments provide are worse than they would otherwise be. Most importantly, it is the very poorest and the vulnerable people who are most severely affected by corruption. Corruption increases the cost of doing business by at least the amount of the bribe paid to secure favorable treatment. Institutionalized bribery also introduces a new set of transaction costs.

In Angola, where diamonds were first discovered in 1912, financial mismanagement was a source of nemesis to majority of the citizens and mine workers and their families.

When the war of independence started in Angola in 1961 Jonas Savimbi financed his guerrilla warfare, with financial resources from sales of diamond. At the end of the 37 years of warring, as many as 300,000 Angolans had died in battle, and hundreds of thousands more had died indirectly. Millions were displaced and the country's infrastructure, seriously underdeveloped at independence in 1975, had been completely destroyed. Two hundred thousand people had been disabled by land mines, more than two thirds of the population lived on less than a dollar a day, and three children out of ten died before their fifth birthday[79].

The world witnessed in Liberia a civil war that was financed by resources from the extractive industry that set the country on the way to total collapse. The cause of the war was financial mismanagement and corruption by government officials. The stage for the war was set by the overthrow of the civilian government of President Tolbert who was thought to be very corrupt in a coup led by Master Sergeant Samuel Doe. The Doe government later became very corrupt and also mismanaged the country's financial resources leading to a campaign to dethrone Doe's administration by Charles Taylor. Before the end of the Liberian civil war about 60,000 to 80,000 people lost their life.

In 2004, the Sierra Leone Truth and Reconciliation Commission (SLTRC) reported that the central cause of the civil war that ravaged the country had been the endemic greed, corruption and nepotism of political elites, who plundered the nation's assets, including its mineral riches… robbed the nation of its dignity and reduced most people to a state of poverty. The citizens of Sierra Leone were denied access to revenue generated from diamonds exploitation. Presidents Siaka Stevens and Joseph Momoh who took over from him were adjudged exceedingly corrupt and tyrannical that all their thoughts centered on how to acquire more personal wealth from the country's diamonds exploitation without recourse to the welfare of the majority of the citizens.

Captain Valentine Strasser, who dethroned Momoh and took over power promised to end corruption. Unfortunately however, Strasser looted the country's diamond resources.

Foday Sankoh who also promised to liberate Sierra Leoneans from the clutches of tyranny however, became a despot who murdered and mutilated the very civilians he said he wanted to liberate. About 75,000 people – most of them civilians – lost their lives in the war.

In the Democratic Republic of Congo (DRC) King Leopold II engaged a private army, the *Force Publique*, which raided and destroyed entire villages, killing men, women and children. Leader after leader such as Mobutu Sese

---

[79] Ibid

Seko, Laurent-Désiré Kabila, Joseph Kabila have all been linked to financial mismanagement and corruption, leading to civil wars and property destruction.

In Nigeria the level of financial mismanagement and corruption cannot be equated with anything as the country has been aptly described as one of the most corrupt nations in the world.

The former Economic and Financial Crimes Commission (EFCC) boss, Mallam Nuhu Ribadu, indicated that as much as US$380 billion has been lost to corruption and waste between 1960 and 1999.

The crises-torn oil-rich communities in the Niger Delta have suffered repression and brutal killings by government security forces in coalition with multinational oil companies, through first, the Rivers state Internal Security Force, comprising the Army, Navy, Mobile Police, Regular Police Force, armed supernumerary Police recruited and trained by the Nigerian Police Force but paid by the Oil Companies as well as armed private security guards; created in response to the Ogoni crises, and later by the Joint Task Force, comprising the Air force, Army and the Navy.

Despite the abundant natural and human resources in the country, Nigerian citizens are faced with grinding poverty as over 70 per cent of the population lives on less than US$1 dollar per day, while at the same time, Nigerians are believed to have on deposit over US$400 billion in foreign bank accounts, most of which were illegally accumulated by those in power entrusted to manage the funds in public interest. This is a clear illustration of the "Paradox of Plenty" occasioned by decades of corruption.

In 1978, Nigeria was ranked the 47th richest country in the world and Nigeria even saw herself as a donor country, establishing the Nigerian Trust Fund at the African Development Bank (AfDB), where it deposited US$100 million to be drawn-on by poor African countries. However, Nigeria was to experience a financial earthquake during the President Shehu Shagari watch, from what many observers of Nigerian political economy believed to have been caused by financial mismanagement and acerbic corruption. An International Monetary Fund (IMF) (2003) report informed that the mismanagement of the oil and gas sector contributed significantly to Nigeria's crippling economic malaise today. According to the report, in 1970 about 19 million Nigerians were poor. However, in 2000, the figure rose to 90 million. Later, the scourge of poverty worsened with about 70 per cent of Nigerians living on less than one dollar a day.

Nigeria has cascaded from being the 47th richest country in the world to its present state of rock-solid wretchedness where everything works in reverse because of corruption.

Over the years, the growth of corruption in Nigeria has become astronomical and this is attributable principally to failure of government to manage social, judicial, political and economic institutional frameworks developed to ensure checks and balances thus leading to break down in laws and policies enacted to ensure accountability and transparency.

The World Bank once reported that as a result of corruption, "Nigeria has lost over US$300 billion in oil revenue, about 50 per cent of projects in Nigeria are dead even before they commence and in fact, the projects are designed to fail because the objective is not to implement them, but to use them as vehicles for looting of the public treasury. 25 per cent of these projects once again according to the World Bank actually get "completed" and "commissioned" with huge fanfare. Unfortunately, most of these projects do not work beyond the day of the commissioning due to a lack of monitoring and continued accountability. Instead of adding value, they become economic drain pipes. These projects have complete structures with boards of directors, management teams, budgetary allocations and capital votes year after year."

The NNPC operated outside the structure of law answerable only to the president or minister of petroleum. For many years NNPC has refused to open its books of accounts for public scrutiny and its operating budget is a mystery known only to its operatives, deducting whatever it considers appropriate as operating cost before remitting what it deems suitable to the Federation Account.

According to the Strategic Union of Professionals for the Advancement of Nigeria (SUPAN) as quoted in The Nation Newspaper of Thursday September 22, 2011 at page 19-20 entitled "Petroleum Subsidy", in 2005 it cost =N=31.50 to produce a litre of petrol but it is sold at =N=65 at a profit of =N=33.50 or 106 per cent. Despite this huge profit it was alleged that due to corrupt practices within the NNPC, this government agency accepted in 2008 that it owed =N=450 billion to Federation account.

Corruption also created the problem of Round tripping –where fuel importers collect subsidy and allegedly ship the same product back to its original place from where it was shipped to Nigeria and then shipped back again to Nigeria and collect another subsidy on the same fuel.

Nigerian government for many years spent trillions of Naira in subsidising petroleum import into this country but the major and independent marketers and some highly placed government officials and politicians and political jobbers allegedly share this money which does not translate to anything for the common Nigerian citizens but continued suffering and abuse of their rights.

In fact, the attempt to remove fuel subsidy in Nigeria has unveiled the magnitude of corruption in the petroleum sector orchestrated by those in high places in this country where trillions of Naira in public funds are swept away into private accounts. This money could have translated to improved quality of life for poverty stricken Nigerian citizens.

Due to corruption, the Nigeria government has been unable to fix its four refineries to functional level away from its current comatose state that has led to fleecing of Nigerians through fuel subsidy on imported or purportedly imported fuel. Building more refineries and fixing the existing ones would abate fuel importation with its consequent corruption.

The House of Representative probe of oil subsidy fund opened the catacombs of corruption in the petroleum sector as did the Honorable Ndudi Elumelu in the electricity sector. Corruption in the aviation industry was exposed following the Dana Air tragedy of June 3, 2012, in which a lot of families lost their loved ones to corruption.

Due to corruption in Nigeria, travelling on Nigerian roads is a nightmare and road safety is a matter of life and death; the retirees are faced with agony because their retirement benefits and pensions are not paid to them as at when due; electricity consumers are inordinately fleeced, there is unprecedented adulteration of products; statutory agencies in this country like Standards Organization of Nigeria (SON), National Agency for Food and Drug Administration and Control (NAFDAC), Consumer Protection Council (CPC) etc. are not well funded to be able to identify manufacturers and importers of substandard products and providers of substandard services in this country.

Many different opportunities exist for bribery and corruption in the extractive industries, each of them with related reputation consequences.

But to proactively engage the triggers to mitigate corruption risks means protecting or even enhancing the company's reputation.

The major critical measures that allow any company to protect and enhance its reputation in the face of a corruption incident are, first, a demonstrable commitment from the Senior Management to engage in ethical business practices and the use of effective established internal set of systems to detect and prevent corruption. The central role of government agencies in overseeing virtually all aspects of the extractive sector lends itself to multiple risks and opportunities for bribery and corruption.

Through the process of government's grant or denial of rights to extractive companies, businesses have an opportunity to improve their reputation by implementing effective and predictable due diligence coordination capacity and execution of protocol that understands the unadulterated role of the government.

Without any level of prevarication there are quite a number of reputation risks associated with corruption at every level of the society but especially in the development and management of the extractive sector that need to be taken into account when addressing issues of due diligence which is a prerequisite to mitigating the risks factors in the extractive sector.

Reputation damages associated with corruption typically arise when a company pretends to have proper anti-corruption measures in place but gets caught in a web of corruption scandal thereby incurring business losses, such as plummeting stock market prices. This is what happened to the French industrial group Alstom in 2014 when it agreed to pay a record $US772.3 million fine for bribing government officials to win power and transportation projects from state-owned entities around the world.

The Petrobras corruption case and its associated reputation damage to the oil and gas giant from corruption scandal that shook the tap roots of the Brazil's state-run oil giant Petrobras amid allegations that former senior executives, construction companies and politicians funnelled kickbacks from hefty oil contracts is a typical example.

Through the police, federal prosecutors and the testimony of former company executives involved in the scheme, Petrobras officials owned up to their involvement in the conspiracy with oil service companies to overcharge for goods and services provided to the company. In this case, the extra revenue said to have been received from the inflated contracts kickback scheme estimated to at least $US800m in bribes and other illegal funds was then kicked back to executives and politicians as bribes and campaign contributions. The corruption scandal resulted in several top executives from some of Brazil's biggest construction and engineering companies to remain jailed while the investigation lasted.

This scandal dealt Petrobras a heavy financial blow leading the company's senior management to cut its investments in 2015 because of its current financial situation and losses resulting from the corruption case. To repair this reputation damage, Petrobras had also to cough out huge financial outlay to hire a compliance officer to head the company's first compliance Program.

To positively rebuild their damaged reputation will require a smart yet long-term strategic approach by the CEO and the board to proactively lead the anti-corruption effort in a broader, holistic manner, and investing in creating a sustainable culture of integrity with all the right policies, incentives and performance metrics in place.

Second, the complex, extensive and diversified oil, and gas and solid minerals supply chains fan the embers of exposure to third-party corruption. To mitigate these risks, companies should implement well-defined policies and establish effective platforms to proactively manage and oversee third parties in all aspects of the business.

This has proven to be realistic where a company actually has proper anti-corruption programs in place and detectors that systematically investigates and report proactively on corruption incidents. In such cases, the company usually succeed in meeting their stakeholders' expectations and enhancing their reputation and business value. This is the case for instance, of Morgan Stanley in 2012 and Ralph Lauren in 2013 that investigated bribery incidents within their companies and voluntarily disclosed their findings to the government.

As a result of corruption, the poorest regions and countries where the extractive companies operate their business seem to be more negatively impacted by the activities of the extractive companies. Therefore, greater global efforts are needed to address and end the corrupt practices of bribery, theft, embezzlement, tax evasion and aggressive tax avoidance of international actors in developing countries where extractive companies operate. Transparency and accountability may pass muster as the most powerful critical tools for natural resource management that will facilitate the end of corruption and costly illicit financial outflows.

6. LACK OF WILL-POWER TO MEET COMMITMENTS

The U.K. based environmental Non-government organization, Greenpeace, accused the British government of reneging its pledge to withdraw its financial support from "dirty energy" projects. The international non-governmental organization alleged that the UK government has reneged on its pledge and dished out more than £1bn in loans for fossil fuel projects around the world.

According to Greenpeace, the UK Export Finance (UKEF) agency dished out a loan of £380m to the Brazilian state-owned energy company Petrobras, Russia's oil giant, Gazprom £430m, as well as £470m to petrochemical companies in Saudi Arabia.

This revelation is inconsistent with a 2010 coalition agreement in which the UK government pledged to move away from fossil fuels in favor of clean energy and a low-carbon economy.

In response to this allegation, A UKEF spokesman responded to the report by saying that the agency would interpret "dirty fossil-fuel energy production" to mean projects producing pollution in excess of international environmental standards, adding that the agency adheres to the OECD Common Approaches, which recommend the undertaking of environmental and social due diligence.

Finally, community relations at the local level is abrasively fraught with human rights, labor rights, and security and corruption issues which in no small measure further expose the extractive sector to grinding risks which can only be mitigated through the development of comprehensive community engagement strategies that work in parallel with their corporate anti-corruption policies, not negating corporate social responsibility framework.

CHAPTER SEVEN

Requirements For Sustainable Extractive Industries Development

An extractive company seeking to embark on an exploitation project activities in a host community must expect some reasonable resistance which is predictable, especially when the local residents and the benefactors of the land where the extractive companies operate are not carried along from the onset. Though the resistance does not need to manifest into violent conflict because with obvious genuine planning the resistance can be effectively managed and reconciled. However, managing community resistance to a constructive result demands skills and patience which seems to be effectively lacking in natural resources sector of the economy.

Companies have made costly mistakes in this area and paid costly prices for such mistakes in the past by rushing to sign agreements with government without really thinking twice of the consequences of side-lining the communities whose land is the subject of the agreement. The extractive companies must have to realize that the government and the communities have a compelling and competing interest on the land and should therefore build a resounding relationship with the host communities early on in the project cycle before entering the host community to start business operations in the first place.

Building such relationships require the companies to invest in a comprehensive baseline study aimed at understanding the economic, social and cultural terrain where the project will be sited to forestall or minimise community agitations. Though this may take time and lot of financial outlay, however, if well conducted it can kick-start a trust-building exercise between and among the parties as well as gather vital information that will put the extractive companies in a position to reap maximum benefits from their long-term investment. Of course this can only be achieved through appropriate community engagement.

This approach has no doubt resulted in positive relations in some jurisdictions in the recent years due to the fact that experts have been investing considerable funds and time to position the business communities and the local host communities to engage in formal dialogue and negotiations. To achieve this also depends on whom to engage in the community critical decision that has to be made early on because engaging the wrong persons would lead to vociferous community opposition.

Secondly, community governance structures and traditional decision-making processes are very complex in many local host communities where women and young people are most often excluded when decisions are taken, yet everyone ought to benefit from the natural resources which are the subject of the decisions and agreements. It is important that decisions regarding natural resources development be jointly taken by the community women, the youths and community leaders.

In engaging the host communities, development companies or their agents need to accord appropriate respect to traditional institutions and the decision-making authorities. The question of how to engage has also in some instances been uncertain and controversial. However, one of the best recommended practices is to listen very carefully, show respect to the community people, not to be in a hurry but allow time for proper engagement with the community, and make presentations in a clear and truthful manner while avoiding the temptation of pre-empting results.

It is not uncommon for some responsible extractive companies to have engaged the host communities in respectful manner in an enriching dialogue with a win-win decision taken by the two parties, only for the host government to overrule the outcome of the dialogue. These companies want to key into doing good things for the host communities, but the government frustrates the effort and the companies get 'burnt'.

Some companies have also taken the hard line position that as long as they have secured government agreement based on national laws, there is no reason to engage their host communities in any dialogue, believing that the combined efforts of the company and government will be enough to neutralize community agitations but some of such companies have had themselves to blame when the host communities rise in agitation against the extractive companies. Nevertheless, despite the position of such companies, they must recognize the need in practice to have broad approval from the host community. Without that, they are setting themselves up for a project that will in all probability fail long-term in the long term.

The secret to succeeding for the extractive companies is finding a path to development where there is genuine mutual benefit for both the local host community and the investor, that is-The WIN-WIN strategy. Therefore, for the project to succeed, the local host communities must see themselves as beneficiaries. As Oxfam America concluded in its 2008 report on "The Costs and Benefits of Mining in Central America", local community support and participation is essential to realising the benefits of investment and reducing the risks of such projects.

The underpinning of this mutual benefit is agreements generally approved which the companies genuinely incorporate into their corporate social responsibilities framework. These agreements could include provision of educational and health facilities, building rural roads, and providing electricity. Revenue-sharing provisions which could generally take the form of a fixed single payment, payment of a per centage of profits or an equity stake in the project can also be factored into negotiated settlements.

Financial disbursement can be agreed to occur in a variety of ways, such as direct cash payment to the local community or through a third-party intermediary like Non-governmental organizations, or deposit into an internal company fund or external trust fund for the community concerned. Providing prospects for economic development and poverty alleviation will also be welcome developments that form fundamental part of the agreements.

Customarily however, the extractive sector hardly provide jobs to host communities members simply because the activities involved in this economic sector are highly technical, most often far above what most of the host communities residents possess. This situation has caused a huge amount of problems for the extractive companies as well as the government and the communities themselves. To avert the problem, it is necessary to divert community agitation regarding their employment by the development companies through appropriate measures that may require the companies to strategically engage the community members through probably the 'local content participation'. This may involve supply contracts or engaging in minor plumbing and electrical works or clearing the project sites and roads. Engaging the local content vehicle means actually treating people in a way that respects their fundamental shortcomings in required technical skill and the same time finding alternative means through which community members can be empowered.

This subtle yet fundamental strategy will no doubt support a long-term friendly relationship between the local host communities and extractive companies that will protect the long-term capital investments. Good community relations are extremely necessary if blood-shed must be avoided and strategic sustainable extractive sector management is to be achieved.

Sustainable management of the extractive sector requires civilised approach at all stages in the extractive sector value chain and this means identification of all the linkages and nexus in the whole system.

Therefore in order to ensure sustainability in extractive sector management, policy-makers must make sure they meet the challenges attendant in the natural resource management ensuring their actions are informed by the understanding of the interconnectedness of administrative systems which must be situated in a political economy context, bearing in mind the constellation of specific local features that will shape the application of civilised approach to natural resources management practice. They also need to be connected to the integrated nature of the institutional frameworks involved in the management of the extractive sector.

The resource-dependent developing countries therefore, need to establish appropriate legal and contractual frameworks; regulation and monitoring of operations; improve on their taxes and royalties collection; revenue management and distribution and sustainable development policies and issues. This approach will no doubt facilitate the transformation of natural resource wealth into wealth for the people, with very minimal collateral damage, in ways that smooth the progress towards achievement of sustainable development.

The new developmental paradigm envisaged with human face in the extractive sector is a developmental plan that appreciates the truth that all individuals have the right and potential to live just; compassionate and dignified lives in sustainable communities without the basic conditions, arrangements and schemes from natural resources development companies and governments that give rise to and perpetuate poverty, injustice, abuse of human rights and dreadful conditions of the environment.

A civilized approach to sustainable extractive sector management, with human face, calls for taking tactical actions to ensure that social, economic, cultural, civil, and political and in fact all human rights of all persons are protected. This is necessary in view of the increasingly globalized world where lives in all manifestations are progressively more interlinked, making it imperative that the new developmental paradigm in the extractive sector must encompass positive changes for all the stakeholders involved; those with power, wealth and influence who control the funds and technology and use more than their fair share of resources; and those most adversely affected by oppressive systems.

The presence of injustices in human society from extractive sector development that traverse slavery, colonization, inequitable trade agreements, corrupt leadership, environmental despoliation that impoverish people, denial of human rights are well recognized in the society, especially in the developing countries and among indigenous communities. The new developmental paradigm without blood but with human face will have no room to continue with these injustices. Where such injustice continues to be noticeable, there must be provision for those affected by the oppressive systems and structures to seek redress in either the home country of the developing company or the home country of the oppressed or abused people.

In the new developmental paradigm in extractive sector development, values and structures that permit over consumption and lack of equitable distribution of available resources must be proscribed.

If blood must be avoided and sustainable natural resources development enshrined instead, all key stakeholders must strategically make the most of the knowledge, skills and resources available to re-orientate and transform the value system of global society towards live and let us live instead of die let us die.

Efforts to promote the new developmental paradigm will involve advocacy work for access to safe water, adequate food, health services and quality education, gender equality, and challenging those with resources, power and influence to effect the needed changes, among others.

The civil and political rights recognize the right of people to self-determination. It is important to adhere to this principle in the new developmental paradigm to promote participation of those people who have been marginalized and whose rights have been violated in all aspects of advocacy work including the joint analysis of their situations, identification of their own needs and assets, contribution of creative solutions, planning and decision-making on the development endeavors to be undertaken and the subsequent implementation, monitoring and evaluation of these endeavors.

Natural resources are increasingly important drivers of sustainable economic growth as it provides governments with revenue to deliver services to their citizens. According to Canada's Sustainable Economic Growth Strategy for International Development, "exports of oil and minerals from Africa, Asia, and Central and South America in 2012 were worth more than US\$1.35 trillion—more than 15 times the amount of official development assistance provided to these same regions in that year. Current trends suggest that the importance of the sector will increase as exploration for minerals, oil, and gas continues in developing countries".

For this obvious reason, developing countries are more and more finding avenues to venture into the enormous revenue potentials held by the extractive sector as a primer to stimulate sustainable economic development and poverty eradication.

The new development paradigm in the extractive sector must therefore improve the governance of natural resources to strengthen national, regional, and local governments and regulatory institutions to manage the extractive sector transparently and responsibly. Governments must establish and implement policies, regulations, and strategies that help citizens benefit from the extraction of natural resources and must ensure improvement in the financial and economic management of their natural resources, through the establishment of sound taxation and revenue-management systems. In addition, acquiring a sound knowledge about their own natural resources sector will ensure the government make sound decisions at all specialized stages of exploration, exploitation, and development.

To ensure sustainable development of natural resources without blood, the host countries should engage the expertise of friendly professionals in the extractive industries to strengthen their capacity for sustainable extractive sector management, so that the natural resources revenues will be better distributed, creating wealth for the people.

More so, the governments, civil society, the host communities and extractive companies must work in partnership. The initiatives pursued through these partnerships will help to reduce social and political tensions, improve stakeholder confidence, and create a more stable and attractive investment climate in the investment jurisdiction.

The new development paradigm will grow other businesses and improve local economic development through support for local economic diversification, leading to foster economic development and improving the contribution of the private sector in host countries by supporting local businesses operating in and around the extractive sector with specific services such as technology, skills development, and financial services so that local businesses can access local, regional, and global markets with better economic potential. Such partnerships between the public and private sector will create better social, economic, and environmental benefits for local communities and long-term value for businesses. Through such initiatives to grow businesses and improve local economic development, there will be reduced economic barriers for women to access employment in and around the extractive sector, particularly in the formal economy.

The internationally recognized Extractive Industries Transparency Initiative (EITI), which seeks to increase transparency in extractive sector has established and implemented some standards on the publication and verification of extractive company tax, royalty payments, and other government revenues. In addition the World Bank Extractive Industries Technical Advisory Facility is a distinctive global venue that facilitates discussions among developing countries on practical issues related to the sustainable management and development of the extractive sector. Also the host countries should take advantage of the Intergovernmental Forum on Mining, Minerals, Metals and Sustainable Development (IGF), which brings together mining officials in developing countries, trading partners, and donors, in addition to civil society organizations and mining companies, to discuss and advance issues of mutual concern, such as international standards and best practices in extractive sector management and governance, to strengthen their capacity to effectively manage their extractive sector to avoid blood-shed.

The host communities should have access to resources for local governments and communities to identify, plan, and manage development projects that address the social and environmental dimensions of the extractive sector and create sustainable economic development PPPs while producing empowered communities and improving public service delivery.

Effective contract negotiation will help to unlock the economic potential of the host developing country, manage their resources more responsibly, reduce poverty, working more closely with the private sector and other partners to advance global development objectives and build tomorrow's markets for trade and investment.

To provide a platform for development of natural resources with human face the problems of poverty eradication, environmental degradation, gender discrimination and good governance must be effectively addressed. The extractive companies, host-country governments and host communities and civil society must partner and promote a framework of sustainable development and establish the mutual trust needed for long-term investments. There is unquestionable need for stakeholders in the extractive sector to develop such trust, in view of the rickety relationships and sustained crises inherent in the extractive sector especially in the developing countries host mining jurisdictions.

If the extractive sector which has provided needed revenues to both development companies and the host countries is to reverse the years of sustained abuse and despoliation for successful investments that ensures that both development companies and host-countries continue to reap the potential benefits of such investments, there must be put into practice a mutually agreeable sustainable investment framework that will last for decades and accommodate unexpected shocks that will arise during the lifetime of the investments.

The stakeholders should be looking at institutionalizing effective framework, focusing on the challenges of governance (e.g. fair and efficient negotiations, contracts, policy and planning framework, the long-term economic relationship of companies and communities, sound resource management), **infrastructure development** (e.g. concession arrangements for shared platforms, corridor development), **economic diversification** (e.g. industrial policy, training, local procurement), **environmental management** (e.g. climate change resilience and adaptation, avoidance and management of catastrophic environmental events), and **integrated rural development** (e.g. promoting economic development at the community and country levels).

Leveraging Public Private Partnerships (PPPs) can also play a major role in building specialized skills required in the extractive sector. The government, private sector, development partners and other stakeholders would need to propose a framework of actions to ensure that the host population reaps a major boost of economic development from its natural resource, while also respecting the profitability of private-sector investments in these important projects.

A cue can be taken from the Sierra Leone's "From Mines to Minds" public private partnership pilot project that brought together the London Mining company; the Sierra Leone Ministry of Education, Science and Technology; the German Agency for International Cooperation (GIZ); and St. Joseph's Training Institute, a nationally accredited and certified training provider. The GIZ's Program document noted that only 23 per cent of middle level and 12 per cent of senior level staff in the mining sector are Sierra Leone nationals, therefore, the project aimed to improve the employment rate of St. Joseph's graduates from the mining district of Port Loko from 40 per cent to 55 per cent; reform the curricula and training delivery for four vocational trades; install infrastructure and equipment for new occupational profiles and ensure efficient use and maintenance of training equipment.

Strengthening the capacity of the host country governments to effectively negotiate large complex projects with foreign investors in the extractive sector would help them to secure the best possible deals in their natural resources development.

In furtherance of the above, natural resources are recognized as public assets in most parts of the world and government is given concessional authority to manage such resources in trust for the people and that being so, the people have the right to know what is being done with their natural wealth.

Transparency, accountability and respect for the rule of law are the central tripod of good governance and the lack thereof of any of these basic elements triggers lack of good governance which has been overwhelmingly recognized as one of the paramount drivers of the deep and agonising weeping, wailing, mourning, and blood-letting prevalent in and around the host communities mostly in developing countries where extractive companies extract natural resource. For this reason, what ought to be a blessing for the people has become their albatross.

With this, the extractive sector has become the citadel of mismanagement of public wealth, poverty, diversion of revenues for personal gain, reduced level of trust in government, communal conflicts, corruption, nepotism, thuggery, maiming, raping, forced relocations, and early deaths among others. All these result from bad governance which is direct opposite of good governance as established above.

Reversing this trend requires the establishment of comprehensible transparent and accountability legal and regulatory frameworks which will increase policy efficiency, reduce opportunities for self-dealing and diversion of public revenues for personal gain, raise the level of public trust and reduce the risk of social conflict. This will further provide the required platform to adequately inform and engage the public and hold the government to account. The overarching goal is to ensure comprehensive transparency and accountability in the management of the extractive sector, from the decision to extract to the granting of concessions, the collection of revenues and the management of resource revenues.

Establishing openness in granting access to natural resources and in the fiscal returns for the state should ensure that the host state receives full benefit from the resource, by attracting the necessary investments to generate financial benefits. A clear legal and regulatory framework will generally serve the interests of the governments and investors better. In addition, transparency helps to ensure that extractive companies operate in non-discriminatory level playing field; reduce opportunities for corruption and may reduce demands from individual investors for special treatment. More broadly, resource decisions involve long-term commitments; reduce the opportunity for the abuse of the citizens and ensuring the confidence and integrity of the resource extraction process.

Transparent rules and regulations for natural resource licenses and concessions should be made available to the public in no ambiguous language, with lucid definitions and explanations of difficult to understand and confusing terminologies. The rules and requirements for access to natural resource, development, fiscal terms, property rights and social and environmental protections requirements, to give citizens a baseline against which to monitor and measure government policies, as well as leveling the playing field for investors will enhance relationships among the key stakeholders.

Fully disclosing the terms of agreements will provide adequate opportunity for legislators and citizens to monitor whether the laws and regulations are being followed and to assess the quality of deals being made on their behalf. This will be in line with the IMF 'Guide on Resource Revenue Transparency' and the Natural Resource Charter which considers the publication of contracts to be best practice. The disclosure should among other areas cover issues relating to the granting of each concession, including public offering documents, lists of pre-qualified companies, successful and unsuccessful bids, contracts and other agreements signed with extractive companies, including the identity of the beneficial owners, resource-related revenues from signature bonuses, royalties, taxes, payments in kind and transit revenues, regular reports showing contributions, earnings, holdings and withdrawals/distributions, including the budget; investment rules, regular independent financial audits related to sovereign wealth/stabilization fund holdings and management; audited accounts for all state-owned extractive companies based on internationally recognized accounting standards, all state-owned extractive companies on a stock exchange and the public agency that has oversight functions on the implementation of the contracts and when their reports should be expected.

Sustainable extractive sector management requires government accountability to an informed public. Resource projects can have significant positive or negative local economic, environmental and social effects, which should be identified, explored, accounted for, mitigated or compensated for at all stages of the project cycle. Alongside disclosure of information, governments should adopt transparent processes for taxing, collecting and managing revenues and for taking spending decisions. Transparency can improve the efficiency and effectiveness of government policies. Public disclosure requirements can improve the quality of data the government gathers and maintains. This makes it easier for relevant bodies such as financial, energy and mining ministries, as well as environmental and regulatory agencies, to do their jobs. Reliable and frequent data can make it easier for governments to plan and manage their budgets and long- term development plans. Transparency also reduces the cost of capital.

Further, there is a need to ensure regular and free participation of legislators, civil society, the host communities, the academia, and the media in the oversight of the extractive sector.

If there is ever going to be a paradigm shift from the unsustainable nature of the extractive sector as experienced by the host communities in the extractive areas in developing countries, to a sustainable approach, the government, industry operators, and the development partners, the United Nations and the Bretton Woods institutions must:

- Provide independent legal advice to the communities to enable the women leaders to weigh options and make informed decisions.
- Provide development advice and build the capacity of women leaders to enable their associations to work effectively together to ensure that the development outcomes specified in their action plans are met.
- Identify and implement strategies to make men better development partners, in concert with women, so that the men are able to play supportive roles in their villages or within their families and can engage in community development policy dialogues and programs.
- Build further capacity to improve women's ability in negotiations, focusing not only on the monetary aspects, but more importantly, how to utilize the money on impactful sustainable projects through proper planning and implementation.
- Develop a communication and awareness-raising strategy to promote and inform the communities, to communicate, educate and raise awareness about all aspects of the extractive sector development, the positive and negative consequences of the development at different stages of development and the impacts as well as what the stakeholders can do to prevent the negative impacts or mitigate them when they occur.

Applying EAP principles internally, without any level of prevarication will:

a) Broaden the scope for dialogue.

The tripartite dialogues are sound initiatives that must be applied to the search for solutions to the numerous and sometimes violent ongoing social conflicts in the extractive sector in all the developing regions of the world where the extractive companies operate.

b) Promote adequate government funding.

It is essential that governments allocate more funds to complete and disseminate precise regulations for consultations with indigenous peoples before issuing oil and mining licenses. These should establish adequate norms for maintaining interactive tripartite dialogues at regional, national and local levels.

c) Encourage a united approach by each government.

Each government will require collaboration between different ministries, sector entities, and provincial and local authorities and should therefore develop strategies that will enable the entities to act collaboratively and effectively in the consultation process under a unified voice, and develop such instruments as inter-agency and stakeholders Memorandum Of Understandings (MOUs) so that divergent viewpoints between ministries (such as for energy and the environment) and the non-state actors can be overcome.

d) Ensure independent consultation.

In order to maintain objectivity and prevent conflicts of interest, the consultation process should be conducted by a government entity that does not belong to the hydrocarbon sector (e.g., the environment ministry or a related agency).

e) Build indigenous capacity.

Capacity building and training programs are badly needed to enable indigenous peoples' organizations to participate actively in the consultation process, as well as in the participatory monitoring of the extractive industries.

The tripartite dialogues are means to exchange information, meet the concerns of the parties, discuss common problems and reach consensus that will serve as a model for improving regulations and practices under which businesses operate and, generally, improve industry contributions to sustainable development and transform natural resources wealth into wealth and quality of life for the people without any encumbrances. In this regard therefore, it is here recommended that:

- The funding sources of the EAP type of intervention should be expanded to embrace mandatory contribution of per centage of profit by the extractive companies; a per centage of the host country's revenue generated from taxes and royalties should be set-aside as contribution towards such initiatives. The United Nations, the developed countries and donor partners should set-aside specific amounts in their annual national or super-national budgetary allocations in order to mitigate funding challenges that will mitigate the bloody extractive sector in order to reverse the rancorous relationship in the sector and achieve sustainability.

- The EAP tripartite experience can and should be considered for other extractive industries that face significant challenges in the region, such as mining, or even for less invasive interventions such as hydropower or wind power. The tripartite dialogues have been proved useful in setting up an agreed-upon working agenda, a common set of principles to govern interactions between the main actors, and a shared set of goals that can contribute to the sustainability of the sector and the wellbeing of local communities.

- This EAP type of Program should be implemented whenever indigenous peoples are affected. Through the tripartite dialogues, indigenous peoples can represent themselves without the intervention of brokers; their voices can be heard directly, and their interests given careful consideration. At the same time, they can be part of a wider discussion on the sustainability of the sector and issues of corporate social responsibility. The dialogues are therefore an effective way to shape interaction between the main actors who represent an opportunity to deal consensually with issues ranging from consultation to compensation, from participatory monitoring to sustainability, and from entrepreneurial partnership to poverty alleviation.

PREVENTING CORRUPTION IN THE EXTRACTIVE SECTOR

Preventing corruption involves eradicating the opportunities that make corrupt tendencies appeal to people and this can be achieved by increasing transparency and oversight of government functions, and improving incentives for good moral character, accountability and transparent performance in public offices as well as following due process in public procurement; providing education and creating awareness, ensuring that citizens understand that corruption is a problem that lowers their standard of living.

Through education citizens can understand the true nature, causes and effects of corruption, acquire improved knowledge of strategies for eradicating corruption and create a knowledge base against corruption. Another strategic approach to eradicating corruption is through advocating against it. An advocate of anticorruption can help to convince fellow citizens and policy-makers that reforms in governance are crucial for economic, social and political development.

The news media and Civil Society Organizations (CSOs) have over the years assumed an importance that could no longer be ignored in governance. The CSOs alongside the media are now regarded as the watchdogs of the government, a role that have been played with significant success over the years. The involvement of CSOs and media in exposing corruption will help bring sanity and reform into the governance process. The roles of CSOs and the media are germane to effective and efficient management of public funds and eventually efficient service delivery and good governance in the country.

Citizens should realize that government belongs to them and must get involved in the act of governance. There is the need to imbibe the principle of participatory governance and decision-making. This can be achieved through participation in the budgetary process; participation in community needs assessments, planning, development, implementation and monitoring of projects in their communities. This participation can be facilitated by the NGOs, Community-based organizations as well as Faith-based organizations.

The government needs to strengthen its anti-corruption institutions to deliver on their mandates. Therefore to checkmate corruption in this regard there must be good governance, increased accountability vide enhanced transparency, effective public expenditure management and financial disclosures, superior oversight mechanism, effective audit, improved law enforcement; have the political will to prosecute corrupt leaders in Nigeria like it was done to the Marcus of Philippines, and in Israel, and other countries in Africa and the world over. The failure of government to bring corrupt officials to book is a pure demonstration of insensitivity to poverty, intractable unemployment and absolute insecurity.

More so, the extractive industry operators need to realize that doing bad business will eventually mean bad business for them and the citizens alike but doing good business means good business for them and host communities and governments. This will ensure proper vigilance and effective countermeasures against corruption.

In furtherance of their activities in the extractive sector, the Bretton Woods institutions need to make some fundamental changes in their goals, structure, operating procedures, policies, and staffing, which are essential to sustain their active participation in the extractive sector without compromising their commitment to infrastructural development and poverty alleviation through effective management of natural wealth on which the developing countries depend upon to alleviate poverty, and strengthen economic development. In line with this, it is here recommended that:

- The World Bank and IMF should ensure that the practical consequences of their operations are consistent with, and do not undermine the goals of the agreements reached at the 1992 United Nations Conference on Environment and Development held in Rio de Janeiro. In keeping with the global commitments made at the Rio Earth Summit, each relevant World Bank project and Program, and each structural adjustment Program of the IMF should ensure that it is consistent with protecting biodiversity and climate, and is in line with international environmental agreements, such as the one protecting the ozone layer.

- The World Bank and IMF should also incorporate into their planning and decision-making processes the value of natural resources and ecosystems to be depleted and/or degraded by policy prescriptions and, in the case of the Bank, the project lending portfolio. For example, natural resource accounting should be incorporated into country programming and loan appraisals.

- World Bank and IMF projects, programs and policies should be reoriented towards local development priorities, as communities are the most appropriate bodies for developing and monitoring local ecosystems. Recognizing that the world's poorest women and men are forced to exploit natural resources in order to survive, self-reliance must be considered a departure point.

- The World Bank must increase its regular loan portfolio to the poor which was a mere US$30 million, which is also less than one-tenth of one per cent of its overall lending portfolio. Out of the US$30 million, the World Bank has already spent US$250 000 on repealing usury laws which protect women from exploitative money lenders.

- In addition, the World Bank and IMF should guarantee the full, informed participation of local communities, cooperatives, small businesses, women's and indigenous peoples' organizations, and other non-governmental organizations (NGOs) in the planning and implementation of their own development.

- An official evaluation of the Global Environment Facility (GEF) pilot phase has found the GEF to follow a top-down, ineffective management by objective approach to dealing with problems related to climate change, biodiversity, international waters and ozone depletion. The World Bank requires that many GEF projects be attached to large regular World Bank loans, which often are at odds with protecting the global environment. The GEF should function independently of the World Bank.

- Full public access to information on GEF and associated projects, guaranteed participation of NGOs and affected communities during the project cycle, and the establishment of an independent secretariat, as well as an effective monitoring and evaluation mechanism for the GEF, are necessary.

- The World Bank should stop funding projects involving forced resettlement for governments and extractive companies that do not have in place policies and legal frameworks that will lead to income restoration for those who will be resettled.

- No project involving forced relocation and involuntary resettlement should also be considered until there is hard evidence that alternatives have been examined, sustainable rehabilitation measures have been put in place in consultation with local communities, and monitoring systems are established that will ensure full compliance with Bank guidelines.

- Specific measures must be taken to hold World Bank staff accountable for violations of the Bank's involuntary resettlement policy. In addition, the Bank should comprehensively provide mitigation and restitution to those people forcibly resettled by Bank projects already underway.

- It is also critical that the Bank, in cooperation with borrowing governments, prepare economic rehabilitation programs for all the populations displaced by World Bank projects since 1980 in violation of the Bank's policy.

- The IMF should guarantee that its programs do not obstruct the goals of civil society to reach equitable and sustainable development committing to the following Program and policy reforms:

a. Required consultation with social and environmental experts for all Fund programs at every stage to ensure that structural adjustment programs do not increase hardship on the poor or aggravate environmental destruction;

b. Information policy that will guarantee increased transparency of IMF policy and programs with full public access to information;

c. Adjustment of the Articles of Agreement to allow greater participation of government ministries and civil society in Program design; and,

d. Creation of an independent evaluation unit to review, on a country-by-country basis, the impact of the implementation of IMF-required or recommended policy prescriptions on poverty, economic development and the natural environment.

BUILDING RESILIENCE IN THE EXTRACTIVE SECTOR HOST AND IMPACTED COMMUNITIES

April 22 of every year has been designated the "World's Earth day" which symbolizes the anniversary of the birth of the modern environmental movement started in the United States of America and European countries especially, dating back to the 1970s.

The need to protect earth's natural resources from the extractive sector development companies amid the climate change threats resulting from continued deforestation, excavations and pollution of the natural environment by the extractive companies continues with increasing urgency especially with the United Nations Food and Agriculture Organization (FAO) enlightenments and disclosures that close to 1.6 billion people or more than 25 per cent of the world's population are dependent on forest resources for their livelihoods.

In September 2015, the United Nations agreed on a set of 17 development goals known as Sustainable Development Goals (SDGs) as the global Agenda for transforming the world by the year 2030. The Sustainable Development Goals (SDGs), are the universal set of goals, targets and indicators that UN members will use to frame their policies over the next 15 years. They are now a centerpiece of international efforts to chart a more responsible course and provide a road map to the universally articulated and accepted sustainable development.

The 17 Sustainable Development Goals with their 169 targets which seek to balance the three dimensions of sustainable development: the economic, social and environmental development has elevated the SDGs to the level of an integrated approach to Sustainable Development which is necessary for the extractive sector to imbibe to ensure their successful operations and protection of the earth and its people.

One critical importance of the SDGs is that each of the 17 SDGs identifies a subject of importance. Specifically, goals 1 and 2 deal with poverty eradication and food security respectively while other goals deal with such issues as gender equality, climate, etc., all of which are critical cross-cutting issues in the extractive sector and in determining whether the extractive sector will honestly contribute to sustainable development especially in their areas of operation.

The provision of "Sustainability Indicators," in the SDGs will help the stakeholders participating in the implementation of the Stakeholders Partnership Arrangement to evaluate and decide the opportunities and challenges of the extractive sector in contributing to reduction in climate change, reduction in poverty in the extractive jurisdictions, reduction in human rights abuses and contributing to the practical achievement of sustainable development.

One of the major importance of integration of economic, social and environmental factors in the SDGs is the need for these developmental components not be pursued independently but through more effective inter-linkages with other entities.

The SDGs integration approach presupposes that the previously held belief of disaggregation has not worked and could alienate distinct interest groups instead of facilitating collaborative achievement of other goals like ensuring viable livelihoods for local populations, environmental protection and respect for human rights.

The United Nations resolve to shift towards more integrated indicators using the SDGs, really holds a lot of promise and a broad relevance that support the view that different aspects of each system are tightly interlinked, which means that striving to improve one indicator may affect others. The clear lesson stemming from this understanding will effectively improve the management of the extractive sector.

It will also positively affect the hundreds of millions of the poor, marginalized or otherwise disadvantaged people who have suffered unusually from the activities of the extractive companies through economic disenfranchisement, rights violations, natural disasters, diseases, conflicts and environmental hazards. Therefore, to ensure sustainable extractive sector management, the stakeholders must reduce this systematic anti sustainable development interlopers. The stakeholders must ensure widespread cooperation to collectively initiate and implement core actions to restructure existing systems of relationships among different stakeholder groups in a way that instils new capabilities in people rather than generating new vulnerabilities thereby, adding to existing human rights abuses, environmental degradation with their consequent insecurity of life and property.

The resilience of host and impacted communities where the extractive companies operate to the adverse impacts of the activities of the extractive companies including global warming, deforestation, erosion and silting, land, air and water pollution and natural disasters such as flooding and tsunamis among others which have left millions of lives destroyed and billions of Dollars in damages along the way have greatly influenced the way they respond to and sustain their livelihoods and well-being thus reinforcing their will to live and not to die.

The level of resilience exhibited by these host and impacted communities ought to trigger the governments and global institutions to work to eliminate the impediments to their freedom to pursue and realize their rights to express their concerns openly, to be heard and to become active participants in influencing their freedom to live a life that one values and to manage one's affairs adequately.

The national governments of the extractive jurisdictions and the global community, especially the developed countries must empower and protect victims of the extractive companies activities by committing to providing them with social services, strengthening their capacity to participate in decisions on how their common natural resources and wealth generated from the resources should be managed, creating employment through local content vehicle among others, to enhancement of their capacity to resist the threats and recover from exposed risks and damages occasioned by the extractive activities of the extractive companies.

Providing access to such basic social services such as education, health care, water supply and sanitation, and public safety can be a powerful force for balancing opportunities and effects; for mitigating and ameliorating sufferings;

providing adequate education to enable the host and impacted communities effectively negotiate with government and extractive companies on rules of engagement to avoid even the ostensible form of rancorous co-existence.

One commonly held misconception is that the host communities and the extractive companies cannot co-exist and live peacefully. This is not the truth because in developed countries there is abundant evidence to the contrary. There are mining companies operating in the US, Canada, UK, and also oil companies operating there without creating the problems witnessed where these extractive companies operate in developing countries. Therefore if it is possible for the extractive companies to peacefully co-exist with their host and impacted communities in developed countries without rancour it must be possible also to co-exist peacefully in developing countries given the right environment to do so such as providing basic social services and social protections because such small early outlay will bring benefits to both the extractive companies and the host and impacted communities that far outweigh the initial small investments.

The case for universal provision of basic social services rests first and foremost on the premise that all humans should be empowered to live the lives they value and that access to certain basic elements of a dignified life ought to be delinked from people's ability to pay. While ways of delivering such services may vary with circumstances and country context, common to all successful experiences is a single idea: The state has the primary responsibility to extend social services to the entire population, in a basic social contract between citizens and state.

Natural and human-made disasters are inevitable, but efforts can be made to mitigate their effects and to accelerate recovery. Providing such social protections as unemployment insurance, pension programs, medical services, potable water and sanitation, can mitigate the hazard and hardship that the host communities suffer from the activities of the extractive sector. These social supports will help households avoid losing their assets, withdrawing their children from school, providing for required medical care, ensuring safeguards during the time of emergency responses and assistance during man-made disasters such as forceful land acquisition without adequate and timely compensation under the government power of eminent domain, destruction of livelihoods support system by the extractive industries, environmental pollution and accidental fires among others, or natural disasters such as floods, erosions, tsunamis, earthquakes, hurricanes and droughts. Emergency response efforts are important and necessary, but building early warning system preparedness and response capacities will ensure resilience and comprehensive positive outcome.

Investment in these social protection services during the early stages of extractive companies' engagement with host and impacted communities of project areas will help bolster the resilience of the host and impacted communities and the economic growth.

Providing employment is a means of sustaining livelihoods, strengthening social connections values, providing for security and welfare of families within the host and impacted communities. Consequently, unemployment conditions are on the reverse related to increase in the rate of criminal activities such as armed and unarmed robbery, rapes, suicide, violence against women and girls, drug abuse and other social problems that can increase personal and communal insecurity. Therefore it is absolutely important that employment generation must be fully integrated into broader policy objectives, like infrastructure development and value chain connectivity such as paying cash for helping the vulnerable poor and unemployed.

Public awareness campaigns and messages aimed at seeking to change people's perceptions and behaviors are essential in ensuring desired social changes that will accommodate the needs of all stakeholder groups. Increasing public awareness and access to information can generate public support for peace and lessen contentious issues that generate and regenerate conflicts.

The stakeholders in the extractive sector must carry everyone along and create policies and institutions that fight exclusion and marginalization, promoting a sense of belonging, trust and offering the opportunity of upward mobility along the communal social ladder to decrease the possibility of conflict conflagration. By using credible and objective intermediaries and mediators, the stakeholders can build trust and confidence among conflicting groups and consensus on issues regarding the sustainable manner to manage extractive industry-related conflicts thereby laying the foundation for the development of 'infrastructures for peace' which help stakeholders to mitigate future crises.

The global framework for universal development through the Sustainable Development Goals approved by the world leaders in September 2015 is a veritable global development architecture that will facilitate cooperation among stakeholders in both developing and developed countries where the extractive companies operate.

All said and done, the following specific actions will be the drivers of sustainable extractive sector management that will move the extractive sector from its present situation of quagmire into a realistic foundation for sustainable socio-economic development.

(a) Empowerment

Empowerment embodies the ideal of individuals and communities overcoming unjust power relations to achieve improved livelihoods and protect their human rights by developing their capacities. The application of the principles of the new developmental paradigm will promote the styles of relationships that strengthen both individuals and community institutions and build technical capacity which promotes empowerment.

(b) Capacity Development

Capacity development implies increasing skills, knowledge and access to resources. The new developmental paradigm envisages the enhancement of the capacity of those group residents of natural resources development jurisdiction whose rights have been violated to overcome poverty and injustice and determine their own future by increasing their skills, knowledge and access to resources. The new developmental paradigm also requires those with resources and power to transform from the culture of over consumption and lack of sharing of available resources to fairness, equity and justice in sharing of available resources.

(c) Non-Discrimination

Non-discrimination honors the dignity of each person and affirms the Universal Declaration of Human Rights. While the new developmental paradigm favors those affected by injustice and oppression, it does not discriminate on any basis including gender, ethnicity, culture, political affiliation, religion, age and sexual identity.

(d) Gender Equity

The new developmental paradigm requires a gender analysis of the roles and relationships of, and between, men and women within the family and broader community. All development has a gender impact and affects women and men differently. The new developmental paradigm requires the participation of both men and women in decision-making and implementation of development activities to ensure that their activities enhance the situation of women and enable both men and women to participate appropriately and promote gender justice relations in the whole community. This includes action to change unequal power relationships between women and men (including ensuring equality of women and men under the law) and that women have access to and control over their share of resources.

(e) Cultural and Spiritual Sensitivity

Development cannot be sustainable unless it embraces respects and enhances and incorporates positive cultural and spiritual practices of individuals and communities into the development process. The new developmental paradigm must recognize this to promote a holistic approach to sustainable development.

(f) Reaffirming Human Rights

Human dignity and well-being are enhanced through ensuring that governments fulfill their responsibility to reaffirm and make effective people's rights: political, social, economic, cultural and environmental. Civil society should continue to play a key role in strengthening those rights and advocating for the accountability of governments and natural resources development companies towards human rights protection.

The new developmental paradigm without blood-letting but human face must recognize the interests and rights of the most marginalized and discriminated groups, addressing issues of access to political and legal empowerment; legal enforcement and access to justice and remedies; and promote organizational policies and procedures that are non-discriminatory but respectful of rights.

(g) Advocacy

Advocacy seeks to address the root causes and effects of poverty and injustice at the community, national and international levels. The new developmental paradigm must promote attitudinal change, mobilize public opinion and strengthen strategic alliances to influence those in positions of power to change oppressive policies and structures in order to promote justice. To be effective, advocacy should start at the grassroots level by facilitating people's ability to advocate for and protect their rights to effectively participate in decisions affecting their lives and livelihoods sources.

(h) Promoting Peace, Reconciliation and Right Relationships

The new sustainable developmental paradigm in the extractive sector must promote a culture of peace and right relationships at all levels, including the home, community, nation and internationally. At a local and global level, it requires an analysis of the underlying causes of conflict and violence and advocates the cessation of oppression, social deprivation and violent confrontation. The new developmental paradigm in communities which have experienced recent conflict resulting from natural resources development must incorporate effective prevention and reconciliation strategies pertinent to such contexts. This will foster and strengthen enabling environment for indigenous capacities and incentives for peace and reconciliation and empowers individuals and groups within societies affected by conflict to cope with past traumatic events.

(i) Effective Communication

Appropriate communication strategy must be advanced in the new developmental paradigm and should start with listening to, respecting and uplifting the voices of those who are marginalized and whose rights have been violated in addition to other key stakeholders. The envisaged communication strategy in the new developmental paradigm must promote communication methodologies which are open and inclusive of gender, race and culture, including honest and transparent communication to those with resources, influence and/or power.

Environmental Sustainability

The new developmental paradigm in natural resources development with human face must challenge policies and practices that impact negatively on the environment. It must work to preserve, maintain and regenerate natural resources through drawing on the knowledge and practices of all people, especially indigenous peoples, and promote the use of appropriate modern technologies.

(j) Over-consumption and lack of sharing

Over-consumption and lack of sharing of available resources by some individuals prevents others from improving their livelihoods and achieving their human rights. The new developmental paradigm must promote sensitization, change in the attitudes and actions of those individuals in the society with excess resources and the ability to effect change and embrace justice for all.

CHAPTER EIGHT

Strategic Approach to Sustainable Extractive Sector Management

It is obvious that extractive companies have been severally accused of all sorts of offenses including human rights abuses, environmental degradation and corruption and also have been successfully sued in the courts of various countries for such offenses. It is also true that some of these companies are still operating and still seeking out other areas endowed with natural resources to launch their extractive activities with their not very human rights and environmentally-friendly records.

Further, it is true that the society is in need of the products and revenues generated from the extractive industries to alleviate poverty, for infrastructural development and general social and economic development. Yet the same society is being relentlessly pummeled by consequent environmental and human rights abuses perpetrated and perpetuated by the same extractive companies who are investing both human and financial resources to ensure successful operations and sustainable economic development.

There is a gamut of human rights abuses and environmental pollution inherent in the extractive sector, and despite these attendant breaches, humanity has also benefitted from the extractive sector development and will continue to do so into the unpredictable future. The extractive sector can therefore be likened to a double-edged sword. In light of this therefore, it is important that proactive steps are taken by stakeholders to mitigate the real and potential collateral damages resulting from the activities of the extractive industries. The proactive steps to be taken must look beyond the current development approaches that have caused the stakeholders to come against each other and instead seek innovative ways to ensure a win-win outcome for all stakeholder groups.

Every man, every woman, and every child on earth, collectively referred to here as man, has either been positively or negatively impacted by the activities of the extractive companies and natural resources exploitation.

Some people have found employment in the extractive sector and some are benefiting from the revenues derived by government and used for development purposes, while some others are end users of the products from the extractive industries. These groups of people may be those positively impacted by the extractive sector.

On the other hand, some people live within the host and impacted communities where the air, land and water are polluted; where their sources of livelihoods have depleted, where their lands are confiscated without compensation, where women are raped, underage children are engaged in extractive activities instead of being in schools, where men have been forced to work without compensation, where people have been forced out of their place of abode into unsettled settlements, where people have been unlawfully arrested and illegally detained and where some others have been extra-judiciarily executed. These groups of people are those negatively impacted by the extractive industries activities.

As a starting point, let everyone both those positively and those negatively affected by the activities of the extractive industries stop for a moment and reflect on the peaceful nature of man when he is asleep. How angry will this person feel if he or she is unceremoniously or abusively waken up from this peaceful sleep? If you are in his or her position, how would you feel? Now let everyone who is positively impacted put themselves in position of those negatively impacted and reflect on his present situation and compare it to when he was positively impacted. Then let those negatively impacted do the same. Then let both those positively impacted groups come and sit together with those negatively impacted groups to decide that there must be a change in the way the extractive sector is managed so that everybody will henceforth become positively impacted by extractive sector development.

Now let these stakeholder groups collaboratively reflect on how many lives, how many ecological fronts, animals, fishes, trees, rivers and streams, agricultural land, mountains and finally humans that might have been saved if they had been meeting and working together on different modalities to ensure a sustainable extractive sector management for all stakeholder groups.

It is not news that one of the sources of the monumental disputes raging among the key stakeholders in the extractive sector is the lack of trust for the extractive companies and host governments by the host and impacted communities, the media and the Non-governmental organizations. These group of stakeholders who see themselves as wearing the halo of victims of the extractive industries activities and the effective champions of the victims of human rights and economic justice respectively, see and portray the extractive companies indiscriminately as wearing the horns of greed and plunder on their head while the extractive companies consequently view them as trouble makers lacking in knowledge and understanding of their extractive activities.

These systematic and structural stereotypes become entrenched by the limited history of constructive engagement among these stakeholder groups who understand little of the divergent camp's terminology, intentions or point of views that creates a deadlock which hardly present an opportunity for dialogue among them. This rancorous relationship is not likely to change or abate any sooner. A breakthrough is therefore needed if consensus should pave and channel the way forward to resolving the often-opposed interest of the extractive companies and people in their environment. This can be achieved if there is concrete evidence that the extractive companies have transparently decided to collaborate with other stakeholders to root out exploitative tendencies to enable them build the needed trust and embrace transparency, while investing in host and impacted communities' development programs.

As the developing countries where extractive companies operate now stand on the brink of full acceptance of the term sustainable development, which has become a watchword, the increasing ability of these developing countries to liaise with supporters from developed countries and outside the extractive jurisdiction in the developing countries to challenge the unsustainable approach of the extractive companies in their business dealings with host and impacted communities without any prevarication will be intensifying the enormous challenges the extractive sector will likely to be confronted with in the near future if they cannot innovate and leverage ascendable collaborative partnerships to improve their business practises, contribute to enhancement of quality of life of the local host and impacted communities while broadening their contribution to the future success of their general business environment.

Progressively minded extractive companies no matter the geographical area and developmental level of the economy where they operate who quickly embrace understanding and adhere to sustainability, linked with the principle of business and human rights are laying the requisite new foundation of a new way of doing business that will enhance other stakeholders value and acceptance for the companies, creating a better future for themselves by optimizing their current performance, as they improve productivity and revenue stream while reducing political risks.

As has already been established, the extractive industries create the most detrimental impact on the environment such as deforestation, erosion, and water and soil contamination among others.

These environmental issues as well as the comparatively large revenue stream accruing through extractive activities, informs the pressure that is generally placed on extractive companies to give back to the earth and society by reducing environmental degradation, working to counterbalance any negative impacts of their operations on the host and impacted communities and maintaining leadership in corporate social responsibility.

In this regard therefore, stakeholders' engagement is especially essential for the survival of the extractive companies as they seek to do business with and extract resources from their host communities. Without effective communication through properly implemented, comprehensive engagement plans and Memorandum of Understanding (MOUs), tension and loss of life and property will continue to be the resultant effect of the relationship. The assurance of workers' health and safety is also a key part of the right corporate strategy as the employees especially those in the field are easily and regularly exposed to precarious situations of the companies' operations and community grievances and bellicosities.

The extractive sector should therefore brace the trend and follow the trail like other business sectors that have realized that in today's world, businesses no longer have that luxury of profit generation and financial maximization for shareholders which used to be their sole operational benchmark as their operational exclusive goals.

The extractive sector operators like other sectors should revolve around how organizations can manage their social and environmental impacts and how they can improve efficiency and natural resource stewardship. This is realistic because more critical stakeholders in the extractive sector the world over since the last few decades, have started to demand for more orientation towards disclosures, sustainability and partnership with host communities as organizations footprint in the society and environment where they operate. Therefore, since it is obvious that value chain enhancement and sustainability are phenomena that have come here to stay, the extractive companies must continuously seek new ways to improve performance, protect reputational assets, and win shareholders and stakeholders trust.

One important thing to understand and agree on is that no one stakeholder group acting alone can provide all the needed solutions and implement all the necessary activities to improve relationship among the stakeholders that will reverse the blood-letting, acrimony and opaqueness in the management of the extractive sector as each stakeholder must individually naturally undertake activities that will maximize its benefits, notwithstanding how other stakeholders are impacted. Therefore to ensure a reversal of the apparent opaque manner in which the extractive sector has been managed would require a comprehensive effort of all key stakeholders working in **partnership** through **Stakeholders Partnership Arrangement (SPA)** framework, which is here recommended as a veritable arrangement through which a sustainable extractive sector management can be universally achieved.

The SPA framework is a gradual, unending improvement activity that involve every key stakeholder in a totally integrated effort that have dramatic effect on meeting or exceeding stakeholders' expectations. It is aimed at improving the quality of stakeholders' relationship by instituting work teams/Total Quality Management approach and equipping the teams with necessary capacity and resources to implement teamwork process and the reorientation of the value system of the stakeholders to ensure understanding of the feelings of other stakeholders leading to eventual sustainable management of the extractive sector.

The SPA is a means of transformation from the functional interests of each stakeholder group to the achievement of a common goal for all. It is a total quality management approach based on team work that must bring to bear the issues concerning *Empowerment, Capacity Development, Non-Discrimination, Gender Equity, Cultural and Spiritual Sensitivity, Reaffirming Human Rights, Advocacy; Promoting Peace, Reconciliation, and Right Relationships; Effective Communication, Environmental Sustainability, Over-consumption and lack of sharing.*

Here the SPA is presented as a methodological participatory management tool which facilitates all key stakeholders involvement in the extractive sector management initiative to work as a team and as partners irrespective of personal interest to achieve a wider success for the wider community of humanity. The SPA envisioned here is a management approach that is planned through participatory processes that facilitate dialogue among different stakeholders and it is aimed to ensure that the previous intervention approaches, some of which were well thought out and designed while some where mere experiments not well thought out or planned, but in all cases both approaches failed to yield positive outcome due to their haphazard approach, will give way to a comprehensive management approach that is well thought out, planned and designed; implemented and monitored and required corrective actions taken by all stakeholders to reverse the colossal failure in transforming natural resources wealth into sustainable wealth for the general public, with the least collateral damage.

With the continuous expanding knowledge of stakeholders on the contributions of the extractive sector to societal development and subsequent negative impacts of their activities especially in the developing world and the quest for sustainable development, there seems to be a general key stakeholders proclivity towards an "*involve me*" philosophy, in which stakeholders increasingly want to be a part of the group contributing ideas that can facilitate active participation of critical stakeholders in the quest for sustainable extractive sector management.

To ensure sustainable extractive sector management, the extractive companies should begin early to engage other key stakeholders to find a "common ground" to build long-term partnerships, develop relationships and propose type and level of stakeholders' engagement.

Once the SPA approach has been agreed by the company as the framework for managing their extractive activities, the relevant stakeholders need to be identified and the extractive companies will then open their doors for effective engagement and interactions with the identified stakeholder group members to generate a portfolio of ideas needed for effective collaboration with heightened positive resultant impacts.

The focus of the collaborative efforts is to build shared understanding among stakeholders by learning from each other to broaden stakeholders' relationships; accentuate improved corporate reputation and community participation; and host and impacted communities resilience to cushion the negative consequences of the extractive sector activities.

The stakeholders who will be involved in the implementation of the SPA must maintain clear, open, continuous, communication that will provide an enabling environment for the interchange of ideas both vertically and horizontally across the stakeholder group spectrum.

This key stakeholder-centered co-leadership-focused approach will enhance the usefulness, the understanding and a sense of unity in the sustainable management of the extractive sector in the sense that the initiative will maximize all stakeholders' advantages by facilitating dialogue between extractive company and its stakeholders and incorporates this input as early as possible in the extractive sector management.

This will also assist the extractive companies in generating needed support and the creation of additional value for the stakeholders that will also guarantee that relevant stakeholder groups are partners and co-travelers in the journey to ensure sustainable extractive sector management. In addition, it is of critical importance because available evidence suggests that there are substantial benefits in this type of collaboration than in the current fragmented extractive sector management approach inadvertently laden with rancorous relationships.

Further, this type of collaboration usually have psychological benefits for the stakeholders such as self-esteem and group-coherence which can improve individual or stakeholder group perception of self-worth, through his or her association with the enlarged stakeholder group and of course the extractive company. The stakeholders in this case, will instead of antagonizing the extractive company which would have ordinarily happened in sole-corporate-management-based approach will see themselves feeling good that they are associating with socially responsible company and also a good corporate citizen.

In the best-case scenario, this approach empowers all key stakeholders to sit and discuss and address problems attendant in the extractive sector that has seen God's natural gift for humanity to develop and improve their well-being turn into an albatross. To succeed therefore, the SPA must be designed and implemented with the active participation of the communities, government, nongovernmental organizations (NGOs), the media, the extractive companies and development partners; who must also be involved in shaping the regulatory and implementation environment in which the required changes will take place.

The SPA is an approach that represents a paradigm shift from many other previous management approaches that have not been successful since they failed to consider and address critical issues necessary to achieve stakeholders' consensus or buy-in and some reasons attributed to these failures, are:

1. Lack of fully engaging communities in the management of the extractive sector.
2. Lack of transparency and accountability in the management of the extractive sector and revenues generated from the sector.
3. Official corruption.
4. Lack of requisite capacity for all stakeholder groups to participate in the management of the extractive sector.
5. Lack of adequate funding to comprehensively engage all stakeholder groups and drive down intervention initiatives to the grassroots.
6. Lack of trust among stakeholders.
7. Gender discrimination, among others.

In dealing with these issues and implementing the SPA framework, it is recommended that the key stakeholder groups focus their approach to embrace the following issues:

1. Land (L): The importance of *land* in decisions regarding extractive sector management and sustainable development.
2. Air (A): The importance of the *air* in decisions regarding extractive sector management and sustainable development.
3. Water (W): The importance of *water* in decisions regarding extractive sector management and sustainable development.
4. You (Y): The importance of *your* contribution as an individual participant in decision decisions regarding extractive sector management and sustainable development.
5. Environment (E): The importance of the *environment* in decisions regarding extractive sector management and sustainable development.
6. Religion (R): The importance of respecting different Religions in decisions regarding extractive sector management and sustainable development.

These issues have been captured under the acronym ***LAWYER.***

Implementation of the SPA Strategic Framework

The stakeholders' partnership arrangement emphasizes the need to put committed people at the center of planning and executing developmental activities in the extractive sector to reduce malicious relationship among the stakeholders and ensure that the extractive sector live up to becoming a catalyst to sustainable social and economic development.

The stakeholders should conceive ideas, develop and implement sustainable infrastructure projects such as roads, jetties, waterways, health, education, employment, industrialization, agriculture and fisheries (aquaculture), housing and urban development and renewal, water supply and sanitation, electricity and telecommunications, where needed in the host and impacted communities.

These infrastructures will stimulate commerce and economic growth and sustainable development in the host and impacted communities which of course will also facilitate sustainable partnership within the key stakeholders groups, leading to robust security, peace and peaceful co-existence and earning the respect and confidence of the host and impacted communities, instead of demanding for such enablers, through various obnoxious channels.

From the above structure, the successful implementation of the SPA strategic framework will involve many individuals performing different activities together and this can make them to be easily distracted, lose track of things and focus. To prevent this from happening requires the leadership of the stakeholder groups to develop a good management system to keep things in order. The stakeholders must do all they can to ensure success, or they will fail in lifting the veil of natural resources curse(s) associated with the extractive sector management.

To ensure successful implementation of the *Stakeholders Partnership Arrangement (SPA)* it is important that leaders of different stakeholders group collaborate to guarantee the implementation of this strategic action is successful by directing and controlling it in a systematic and transparent manner. This can be achieved by using International Organizations for Standardization (ISO) 9000 series Quality Management System (QMS) designed for it.

Implementing this collaborative effort through the ISO 9000 series QMS framework is not only strategic, it is indispensable because it is an internationally approved requirement for managing organizations around the world and it is also the internationally accepted QMS Model to follow in setting up and operating QMS.

In addition, ISO 9000 QMS is developed around the 8 Quality Management Principles set forth below:

Principle 1: Customer Focus

Principle 2: Leadership

Principle 3: Involvement of People

Principle 4: Process Approach

Principle 5: System Approach

Principle 6: Continual Improvement

Principle 7: Factual Approach to Decision Making

Principle 8: Mutually Beneficial Supplier Relationship

Principle 1: Customer Focus

This implies understanding current and future customer needs; meeting and striving to exceed customer expectations. In this case, the customers are the critical stakeholders in the extractive sector which has already been established in chapter two of this book. This is so because they are the main beneficiaries of effective extractive sector development.

Principle 2: Leadership

The ISO 9000 QMS mandates that leaders should establish unity of purpose and direction of the organization; create and maintain the internal environment in which people can become fully involved in achieving the organization's objectives. This no doubt will help the leadership of the stakeholder groups to deploy innovations to achieve sustainable extractive sector development.

Principle 3: Involvement of People

All critical stakeholder groups at all levels must fully participate in the process as their full involvement enables their abilities to be used for the benefit of all.

Principle 4: Process Approach

The ISO 9000 QMS requires that a process which is a set of interrelated or interacting activities which identifies processes, inputs and outputs, interactions, transformation of inputs into outputs, measurement criteria, continual process improvement that produces desired result is more efficiently achieved when resources and activities are managed as a process. This will be a necessary requirement for the implementation of the *SPA* as a strategy to lift the natural resources curse and guarantee sustainable development through the extractive industries.

Principle 5: System Approach

Again, the ISO 9000 QMS requires identification, understanding and managing a set of interrelated processes as a system which contributes to effectiveness and efficiency in achieving objectives. The stakeholder groups must understand the dynamics of a system as a set of interrelated processes, requiring that the output of one process is an indispensable input to one or more subsequent processes, and so on. It is therefore of critical importance to manage the interface between processes very pragmatically to ensure that the overall system is efficient and effective.

Principle 6: Continual Process Improvement

Further, ISO 9000 QMS recognizes continual process improvement of overall performance as a permanent objective and this must be a planned activity.

Principle 7: Factual Approach to Decision Making

Furthermore, ISO 9000 QMS recognizes that effective decisions are based on the logical and intuitive analysis of data and information. Therefore, it is important the stakeholder groups imbibe decision making process to ensure success in the implementation of the SPA strategy as essential innovative tool for sustainable extractive sector management.

Principle 8: Mutually Beneficial Relationship

Mutual beneficial relationships among the stakeholders will enhance the ability to create value. In this strategy, the stakeholders are sensitised on the need to work together, bringing to their knowledge, the benefits of collaboration and effective communication as well as respect for the rights of others.

The stakeholders will be grouped into the following work-teams:

1. The Senior Leadership Forum
2. The Consultation Team
3. The Process Improvement Team
4. The Monitoring and Evaluation Team

The Senior Leadership Forum

This forum should be comprised of the senior leadership of the entire stakeholder's group which include the government, the development partners, the extractive companies, the host communities, the civil society, the media, and the employees' unions. The Senior Leadership Forum is the highest decision-making organ of the stakeholders' working group. Their decision which must be reached through consensus on issues affecting the extractive sector is final.

The government representatives in this team should be in the rank of ministers or secretary (where secretary is used to designate such a high position as in the US), who have the authorisation of the executive president of the country to negotiate agreements and enter into binding contracts on behalf of the government. The government must respect and implement decision and agreement entered on its behalf in the spirit of *Pacta sunt servanda*.

The extractive companies should be represented by their highest ranking member such as the chairman or president of the company in the country of operations. These people are well positioned to take binding decisions for and on behalf of the companies which they represent.

The development partners should be represented by their respective country directors who have the competence to take binding decisions and contract on behalf of the development partners. For instance, the WB is expected to be represented by the Bank's country director in the country of operations.

The host communities should be represented by their leadership and of course women must be part of the constituent representatives. This is imperative as women who play leading roles in development and sustainability of humanity bear the brunt of the fallout of the activities of the extractive industries. Their perspective on development must not be foreclosed. They are critical stakeholders whose genuine contributions will contribute significantly to improved extractive sector management.

The Civil Society Organizations should likewise be represented in this forum by the heads of the organizations such as the Executive Directors or Presidents as the case may be. The media should also be represented by the most senior members of the media outfit. Being that the Editors are the most likely people available with the relevant decision-making powers, except the Chairpersons and Directors, the Editors, the Chairpersons or the Directors can represent the media houses.

The employees' unions will be represented by the leadership of the union. The participation of the union in this forum is important because the staff work in the field, meet on daily basis with communities members and witness first-hand, the impacts of their activities on the host communities and also the employees are the first targets of community agitation, who also suffer the fallout of industrial accidents.

This team should embrace fully the commitment that goes beyond rhetoric and set the standard of operation through the example of their performance on how other teams can function. The success of the SPA depends on their leadership quality. They must ensure they are fully committed and take decisions that benefit the entire stakeholder groups instead of just merely looking out for the interests of the stakeholder group they represent. The major objective of this team is to direct the overall progressive agenda of the SPA to ensure sustainable development in the extractive sector where positive impacts of the extractive sector is considered more important; and a shift away from the operations decisions based only on maximizing profits.

This team should work closely with the consultation team who are the egg-heads that will conduct research, collect and analyze data and come up with recommendations as to how to improve the network of relationship among the stakeholders groups to ensure mutually agreed sustainable development framework.

The Roles of Senior Leadership Forum

The roles of the Senior Leadership Forum will be to:

a. Establish the quality SPA implementation policies and objectives;
b. Promote the established quality SPA policies and objectives;
c. Ensure focus on stakeholders requirements;
d. Ensure approved processes are implemented;
e. Ensure QMS is in place;
f. Ensure the availability of resources;
g. Review the QMS periodically;

h. Decide on actions on policies and objectives;
i. Decide on actions to improve QMS.

The Consultation Team

The consultation team will comprise the multidisciplinary team of experts (Eggheads) from the government, extractive companies, the CSOs, the media and communities. They can include professionals like medical doctors, nurses, engineers, lawyers, accountants, business managers, psychologists, sociologists, agriculturists and environmentalists among others. This team will conduct researches, analyze data and come up with best possible solutions to ensure every stakeholder interest is appropriately factored into developmental decisions.

This team will focus their actions on ensuring the economic, social and environmental wellbeing of the entire stakeholder groups. Their work should focus on how to ensure sustained profit for the investors, how to ensure the rights of the community members are protected, that the environment is properly managed with little collateral damage; that the cultural heritage of the host communities are upheld and that the extractive sector development is beneficial to all stakeholder groups. They will also ensure that through revenues generated from the extractive sector there will be improved infrastructural development, improved standard of living, quality education and healthcare delivery services and improved security.

This team advises and recommends to the senior leaders' forum, the best possible actions to take that will benefit if not all, most of the stakeholders at once. This team also designs appropriate strategy in collaboration with process improvement team to ensure effective implementation of the recommended activities and actions.

The Process Improvement Team

The extractive sector has the potential to generate significant wealth that will put permanent epitaph of happiness on the faces of the poor residents of the host communities and host developing countries where extractive operations take place; and also serve as vital mechanism for sustainable growth and poverty alleviation through the huge revenues generated from royalties, taxation, and exports and employment. This will make the extractive sector a catalyst for the transformation of life; driving of economic growth, creation of jobs and reduction of poverty in resource-rich developing countries. Too often, however, these opportunities are missed as the extractive companies do not mostly ensure these opportunities materialize and the in the end what follows is that the extractive industries deliver as much or more damage than benefits.

In the extractive sector major stakeholders visibly present that influence or are influenced by activities of the extractive industries, or other stakeholders whether positively or otherwise, have been identified to include the government, the development partners, the extractive companies, the host communities, the civil society, the media, and the employees union.

Also in many developing countries of the world where the extractive companies operate, there have been tales of woeful environmental despoliation and fatigue, bionetwork destabilization, and human rights abuses that consist of unlawful arrests and illegal incarcerations, forced-labor, forced-displacements or arbitrary relocation of residents without adequate and timely compensation, rapes and extra-judicial killings, meted out to local residents of these natural resources development jurisdictions. These have created a vicious cycle of relationship among the stakeholders.

To reverse the seemingly intractable vicious relationship existing among stakeholders in the extractive sector will require direct and active participation of key stakeholders in a teamwork process that will smoothen the rough edges of the relationship that exists among the stakeholders. This process will promote increased awareness and understanding of extractive industries-related issues; it will promote learning, and trigger appropriate actions among the stakeholders.

This book ultimately aims to improve the management of the natural resource endowment of developing countries in such a manner as to make sure that the revenues generated from natural resource extraction contribute optimally to achieving sustainable economic growth and livelihoods enhancement and reversing the age-long agonies of the host and impacted communities and states by identifying and analyzing the historical challenges overwhelmingly present in the extractive sector.

Improving relationship among stakeholders requires the stakeholders coming together and agreeing to work together to improve their not so friendly relationship by every stakeholder group coming together and laying their concerns on the table for discussion with a view to resolving them in a consensus manner. The process improvement team will be saddled with the responsibilities of ensuring that there is improved relationship among the stakeholders. This team will comprise of experts in conflict resolutions and peace building, team building practitioners, relationship building practitioners among others.

The goal of this team is to continuously improve relationship among the stakeholders. The team will be responsible for providing training to the stakeholders that will build their capacity to understand rights and responsibilities towards ensuring effective and efficient management of the extractive sector.

The Monitoring and Evaluation Team

One of the accusations constantly levelled against the extractive companies is that they always renege on promises such as agreements reached between the host communities on companies' corporate social responsibilities of developing and providing infrastructures such as roads, healthcare centers, school buildings, potable water and electricity and community youths employment. This has on numerous occasions activated communities' agitations against the extractive companies.

By reneging on agreements reached with host communities, the extractive companies have knowingly or unknowingly exposing themselves to political risks such as vandalism of extractive companies' equipment, kidnapping of staff, demonstrations against the extractive companies and blockage of entrances, thereby shutting the staff out of their offices and extractive areas. In response, the extractive companies in collusion with host states have responded to these actions of the host communities in such a way that has violated the fundamental rights of the host communities.

To avoid this type of unhealthy relationship, there is need for all the stakeholders to ensure that they carry out agreed responsibilities. This is where the monitoring and evaluation team comes to play their roles.

This team will make certain that all activities recommended by the consultation team and approved by the senior leadership team for implementation are carried out with utmost dexterousness. They will report their findings to the senior leadership team who has the authority to enforce compliance.

Their monitoring exercise must be transparent and conducted in an atmosphere where information about their activities can be communicated without corporate pre-screening.

Benefits of Using ISO 9000 to Stakeholders in Implementing the SPA

The adoption of ISO 9000 QMS by the stakeholders in the extractive industries in the implementation of the SPA strategic framework will lead to:

- Increased stakeholders' morale as they are united for a common task and are empowered to take control of their assigned processes.
- Reduction in human rights abuses and environmental pollution, degradation and despoliation.

- Reduction in conflicts among the stakeholders.
- Improved communications among the stakeholders.
- Improved relationships among the key stakeholder groups.
- Improved and sustainable livelihoods and quality of life for host and impacted communities.
- Reduced tension and infrastructure destruction.
- Increased revenues for the industry, government and shareholders.
- Reduced gender inequality, among others.

Obstacles to Implementation of the SPA Framework

Inadequate capacity on the part of some of the stakeholders to effectively engage each other could become an obstacle to their successful implementation, monitoring and management of obligations and commitments under this SPA framework. The problems could be compounded when the various stakeholder groups and signatories to the arrangement fail to understand their roles and responsibilities in the SPA framework or engage in contradictory and conflicting acts that could hamper transparency in resolving identified issues and implementing agreed solutions.

Ineffective coordination of this arrangement will pose a serious challenge to achieving sustainable extractive sector management through the SPA dispute resolution mechanism. If not well coordinated and line responsibilities delineated and respected by all stakeholder groups then there will be disagreement on the strategy to pursue which is likely to have a negative impact on resolution of disputes.

Lack of sufficient knowledge on pertinent procedural rules or on the line of action to pursue, could encumber the process of dispute resolution and enhanced stakeholders' relationship.

The above mentioned impediments, when compounded by other factors, could lead to a point of crisis.

Funding Sources to Implement the SPA Framework

To ensure successful implementation of this recommendation requires financial resources. It is therefore recommended here that such funds will come from the extractive companies, the host states, development partners and donor agencies.

The extractive companies can set aside a per centage of their annual profit or can factor their contributions into their annual operating budget. The host governments can also set aside a per centage of their revenues from the extractive sector or through annual national budgetary allocations towards the implementation of this stakeholders' partnership agreement.

However, where the extractive companies operate a joint venture partnership, the operating cost of the SPA could automatically be built into the project budget.

The development partners and donor agencies can raise funds from their supporting foundations or funding sources to contribute to the process.

If the SPA strategy framework outlined above is well implemented and some of the more frequently observed issues and challenges that have been highlighted in this book as militating against sustainable extractive sector management are adequately addressed and resolved as recommended, at the end of the day, the result will be "*Sustainable Extractive Sector Management*".

BIBLIOGRAPHY

Constitution, Treaties and Regulations:

- Constitution of the Federal Republic of Nigeria, 1999, as Amended.
- MIGA Convention, Art 11(b) and (c)
- UNCTC Regulations.
- UNCTC, *Transnational Corporations in World Development: A Re-Examination*. E/C.10/38 of 20 March 1978.
- UN Global Compact July 2000.
- United Nations Guiding Principles on Business and Human Rights 20115.
- UN Resolution 626 (VII) of 2 December 1952.
- Venezuela Constitution (1961) as amended.
- Vienna Convention on the Law of Treaties (1969).

Professional Publications (Books):

- Adam Hochschild. (1998) *King Leopold's Ghost* Boston: Houghton Mifflin.
- Arsel, M. (2005) *'Risking Development or Development Risks: Probing the Environmental Dilemmas of Turkish Modernization'* PhD Dissertation, University of Cambridge.
- Austin Onuoha. (2012) *the State and Status of Human Rights and Business in Nigeria:* Research Report of Africa Center for Corporate Responsibility (ACCR).
- Basil Davidson. (1973) *In the Eye of the Storm: Angola's People,* Garden City, N.Y: Anchor Books.
- Boniwe Igbadi Z. ((1999) *Do Bilateral Investment Treaties (BITs) Contribute To Customary International Law If so, how?* – A Master of Law (LL.M) Dissertation submitted to the Center for Energy, Petroleum and Mineral Law and Policy; Faculty of Law and Accountancy, University of Dundee.
- Boniwe Igbadi Z. (1998) *Sovereignty over Natural Resources on Modern Petroleum Contract.*

- Boniwe Igbadi Z. (2004) *Towards Sustainable Development in the Solid Minerals, Oil, Gas and Energy Sectors,* JEU ACE Publishers, Lagos.

- Boniwe Igbadi Z. (2007) *Earths Natural Resources: A New Paradigm for Sustainable Development* A Publication of the Center for Sustainable Socio-Economic Development.

- Bradney, A. et al. (1995) *How to Study Law* 2nd. Ed. Sweet & Maxwell.

- Cassese, A. (1994) *International Law in a Divided World*, Clarendon Press, Oxford.

- Christos N. Pitelis and Roger Sugden (Editors). (1991) *the Nature of the Transnational Firm*, Routledge, London.

- Coban, A. (2003) "*Community-based Ecological Resistance: The Bergama Movement in Turkey.*" Environmental Politics 13(2).

- Cruz, Wilfrido and Robert Repetto, 1992 "*The Environmental Effects of Stabilization and Structural Adjustment Programs: the Philippines case*", Washington, DC: World Resources Institute.

- Danilenko, G. M. (1993) *Law-Making in the International Community*, Martinus Nijhoff Publishers, Dordrecht.

- Esa Paasivirta. (1990) *Participation of States in International Contracts and Arbitral Settlement of Disputes,* Finnish Lawyers' Publishing Company, Helsinki.

- George Schwarzenberger. (1969) *Foreign Investments and International Law,* Stevens and Sons Ltd, London.

- Hans VanHoutte. (1995) *the Law of International Trade,* Sweet and Maxwell, London.

- Harris, D. J. (1998) *Cases and Materials on International Law*, 5th ed. Sweet & Maxwell, London.

- Henry Campbell Black. (1995) *Black's Law Dictionary*, 6th ed. West Publishing Co.

- Ian Smillie. (2010) *Blood on the Stone: Greed, Corruption and War in The Global Diamond Trade* London: Anthem Press.

- Ignaz Seidl-Hohenveldern. (1992) *International Economics Law*, 2nd revised ed., Martinus Nijhoff Publishers, Dordrecht.

- International Business Leaders Forum, (2008) *EITI Business Guide: How companies can support implementation.*

- Kolo, A. (1994) *Managing Political Risk in Transnational Investment Contracts* (Center for Energy Petroleum and Mineral Law and Policy: Dundee, Scotland.

- Les Roberts *et al. (*2003) *Mortality in the DRC: Results from a Nationwide Survey* New York: International Rescue Committee.

- Magdalena M. & Martin Martinez. (1996) *National Sovereignty and International Organizations*, (Kulwer Law International, Hague.

- Martin Dixon. (1996) *Textbook on International Law*, 3rd ed. Blackstone Press Ltd., London.

- Martin Dixon & Robert McCorquodale. (1991) *Cases and Materials on International Law*, 2nd ed. Blackstone Press Ltd., London.
- Owusu, J. Henry, 1998. "*Current Convenience, Desperate Deforestation: Ghana's Adjustment Program and the Forestry Sector*" Professional Geographer 50 (4).
- Özen, H. (2009) *'Located Locally, Disseminated Nationally: the Bergama Movement'*, Environmental Politics, 18: 3.
- Passmore M. Hamukoma. (2011) *Survey and Analysis of Demand for and Supply of Skilled Workers in the Zambian: Mining Industry.*
- Patrinos HA, Barrera-Osorio F, Guáqueta J. (2009) *the Role and Impact of Public-Private Partnerships in Education,* World Bank.
- Philip Wood. (1980) Law *and Practice of International Finance*, Sweet and Maxwell, London.
- Richard Fanthorpe and Christopher Gabelle (2013) "*Political Economy of Extractives Governance in Sierra Leone*" (World Bank Publication No. 83425).
- Robert Pritchard (editor). (1996) *Economic Development, Foreign Investment and the Law*, Kulwer Law International, London.
- Shihata I. (1988) *the MIGA and Foreign Investment.*
- Shihata I. (1992) *World Bank Legal Framework for the Treatment of Foreign Investmen*t, Vols. 1 and 2.
- Sornarajah, M. (1995) *the International Law on Foreign Investment*, Groutius Publication Cambridge University Press.
- Steer, Andrew, 1996. "Discussant's Comments" in Gandhi, Ved P., ed. *Macroeconomics and the Environment* Washington D.C.: International Monetary Fund, 1996, 65-68.
- The World Bank (Extractive Industries Transparency Initiative – Multi-Donor Trust Fund (EITI MDTF) Annual Report 2013) *Building on Progress to Implement the EITI Standard.*
- Timothy Hillier. (1994) *Public International Law-Lecture Note Series*, Cavendish Publishing Ltd.
- United Nations. (2013) *The rise of the south: human progress in a diverse world*, Human Development Report.
- Wolfgang Friedmann. (1964) *the Changing Structure of International Law.*
- World Bank (2013) *Building on Progress to Implement the EITI Standard*, EITI MDTF 2013 ANNUAL REPORT No. 83538.
- World Trade Organization (WTO) (2005) *Trade Policy Review: Republic of Guinea – report by the Secretariat (revision)*, Report No. WT/TPR/S/153/Rev.1, Geneva: Trade Policy Review Body, December 14.
- Yasap Popoitai and Waafas Ofosu-Amaah (December 2013) *negotiating with the PNG Mining Industry for Women's Access to Resources and Voice: The Ok Tedi Mine Life Extension Negotiations for Mine Benefit Packages* A publication of the World Bank Institute: Publication No. 84838 V1&2.

Articles and Magazines

- Arjun, Goswani. (October 1997) *Risk Profiles in Project Financing*, International Business Lawyer Vol. 25 No. 9, P. 406-409.
- Bruce Johnson and David Nanson. (October 1997)*Power projects in India*, International Business Lawyer Vol. 25 No. 9, and P. 395-399.
- Dholakia, Lord and Phillips, Lord of Sudbury, Sir Nigel Rodley, Kirsty Brimelow QC - Chair of Bar Human Rights Committee.
- Gwynne Skinner, Robert McCorquodale, Olivier De Schutter & Andie Lambe 01 Dec 2013.
- Hanson, P. and Aranda, V. (1991) *An Emerging International Framework for Transnational Corporations*, 14 Fordham *ILJ* P. 881.
- *Issues Paper. (2013) A Coordinated Approach to Mineral Skills Development in Africa* African Mineral Development Center.
- Lauren Carasik. (April 22, 2014)*Organization's Investor Protection Panel Dis-empowers Marginalized Communities* Aljazeera News.
- Lipstein, K. (1945)*The Calvo Clause in International Law*, 24 *BYIL* P. 130.
- Maura McGowan QC - Chairman of the Bar, Phil Lynch - Director, International Service for Human Rights & 12 other lawyers 02 Dec 2013.
- Nick Fluck - President of Law Society, Martyn Day Leigh Day, Carla Ferstman - Director of REDRESS.
- Olatunji Dare. (February 11, 2014) *Corruption: The EU to the rescue* The Nigerian Nations Newspaper Tuesday.
- Oyejide, A. and A. Adewuyi. (2011) *Enhancing Linkages of Oil and Gas Industry in the Nigerian Economy*, MMCP Discussion Paper 8.
- Shihata I. (1986) *Towards a Depoliticisation of Foreign Investment Disputes: The Role of ICSID and MIGA*, *ICSID* Rev. 1.
- The Economist, Why Iran and America Clash, (April 11th 1998).
- Tobi, N. (1991) Legal *Aspects of Foreign Investment and Financing Energy Products in Nigeria*, 14 Dalhousie *LJ* P. 5.
- Redfern, A. (1984) *The Arbitration between the Government of Kuwait and Aminoil*, 55 *BYIL* P. 65.
- Ryan Lizza. (24 July 2000) *Where Angels Fear to Tread, The New Republic*.
- Zimman, Jenna E. (1998) *Freeport-McMoran, mining corporate greed*. Z 0 Magazine http://www.lol.shareworld.cxom/zmag/article/jan98ziman.htm

Cases:

- *AETNA Casualty and Surety Company v Bramwell*, 12 F. 2d 307, 309.
- Aminoil Case (1982) 66 *ILR* 588.
- *Buck v Attorney General*, (1965) Ch. 745.
- *Czech Republic v CME B.V*, Review by Svea Court of Appeal, 15 May 2003, 42 *ILM* 919 (2003).
- *Columbia v Peru I.C.J* Reports 1950.
- *De Haber v the Queen of Portugal*, [1851] 17 Q.B 171, 207.
- I Congreso del Partido, [1981] 2 ALL ER 1064.
- *Iran v United States*, Claim Tribunal (27 March 1986).
- *Kuwait Airways Corporation v Iraqi Airways Company*, (1995) 1 *WLR* 1147.
- *LIAMCO v Libya Arab Republic*, (1977) 20 *ILM* 1.
- *Mobil Oil v Iran*, 16 Iran-U.S. C.T.R 3; 40-43.
- *Nicaragua v US, I.C.J* Reports 1984 p. 392
- *Republic of Ghana v Telekom Malaysia Berhard*, District Court of The Hague, 18 October 2004
- *Schooner Exchange v McFaddon*, (1812) 7 Cranch 116.
- *Texaco v Libya Aral Republic* (1977) 53 *I.L.R.* 389, (1978) 17 *I.L.M.* 1
- *Trendtex Trading Corporation v Central Bank of Nigeria,* (1977) 1 ALL ER 881.
- *Victory Transport Inc. v Comisaria General De Abastecimientos y Transportos,* 25 *ILR* 110 (1963).

ADDENDUM

ADDENDUM

Case Study
Management Institute Initiative (M-II) Versus Basil Community Development Association (BCDA)

"There can be no friendship without confidence and no confidence without integrity".

Samuel Johnson

This case study is an account of the relationship between Management Institute Initiative (M-II) and Basil Community Development Association (BCDA) in an effort to ensure sustainable extractive sector management in Basil Community which is a mining host community. This case study merely serve as an illustration of how well-conceived extractive sector sustainable management initiative can be rubbished, mismanaged or turned into conflict and crises-breeding ground either through lack of focus and vision; proclivity for greed, corruption, deception, fraud and nepotism; incompetent management or due to lack of requisite leadership skills or outright sabotage and Conflict Entrepreneurship.

Establishment of BCDA

When Corporate Extractive Company Developers (*CECD*) which is a mining company operating in Basil Community experienced an increased open agitation and sustained violent confrontation against their operations, they understood the urgent need to engage their host community to collaborate in an effort to implement CECD Corporate Social Responsibility framework aimed to mitigate the open agitation and sustained violent confrontations against their operations to achieve sustainable extractive sector management, they approached the leaders of Basil Community which is the host mining community and informed them of their plans to reduce tension between CECD and the host community.

To set the plan rolling, CECD established a corporate entity called Management Institute Initiative (M-II) that will engage Basil Community and the agreement reached with Basil Community leadership was that Basil

Community will establish a Basil Community Development Association (BCDA) that will work with M-II in a partnership arrangement to ensure effective mitigation of escalating conflict between the company and Basil Community. However, the funding for the operation of the BCDA in Basil Community will be provided by CECD. This fund will be channelled to BCDA through M-II. The implementation strategy was that M-II and BCDA will be independent of each other but M-II will be responsible to facilitate registration of BCDA to give the latter a legal backing.

The intended Results are that:

A. *Conflict issues in the Basil Community are effectively managed through the collaboration and collective action of a broad range of individuals, businesses and organizations.*

bB. *Emerging conflicts are identified and mitigated through agreed systems in Basil Community through the BCDA.*

C. *Basil Community will mobilize individuals, businesses and organizations from Basil Community who share a common interest in addressing the issues causing the agitations and conflicts between CECD and Basil Community.*

Following this agreement, Basil Community elected the members of the community who will represent the community in BCDA.

The first mistake made in the incorporation of BCDA as an independent corporate entity was that Members of Board of Trustees who incorporated the BCDA were employees of M-II, instead of representatives of Basil Community.

Upon the incorporation of BCDA as an independent organization, CECD took the initial step to position the organization to become a credible organization representing the Basil Community. CECD provided agreed funds and sent in their highly experienced staff to guide the process and seconded another staff an internationally recognized civil society organization on permanent basis to head the M-II Team who will directly work with the BCDA to midwife the transition process to enable BCDA manage their affairs independent of M-II.

M-II Partnership with BCDA

When M-II approached Basil Community and requested for collaboration to prosecute the agreed CECD Corporate Social Responsibility initiative in Basil Community, M-II came with the idea of partnership and project ownership by the extractive host Basil Community, not ownership by M-II.

The BCDA was inaugurated in the Basil Community with their Executive Committees (EXCO) to represent Basil Community.

M-II Collaboration with BCDA

M-II asked BCDA to contribute their quotas as partners in every project activity to be carried out in Basil Community by providing workshop venues, public address system, engaging other critical stakeholders, sending out invitations to stakeholders, and sometimes providing and paying the resource persons that will be invited to facilitate required trainings.

Adoption of Agreed CECD Byelaws

The Byelaws that will guide the operation of BCDA was drafted in collaboration with M-II. The Byelaws was adopted by the leadership of Basil Community; providing opportunity for M-II Team to transfer the funds earmarked for BCDA activities to BCDA.

Inauguration of the BCDA leadership

The BCDA leadership which will be responsible for the facilitation, coordination and shaping the strategic direction of the BCDA was inaugurated on December 20, 2012. Thus the management framework for the operation of the BCDA was fully established.

Issue No 1: Transfer of Leadership to BCDA

After the inauguration of the BCDA, M-II management refused to transfer earmarked BCDA funds to the BCDA elected executives. Many observers and some of those who worked for M-II who were very instrumental in the conceptualization of this project and subsequent design of implementation strategy, and have resigned for various reasons, believe that the refusal of M-II management to transfer the funds and allow BCDA to be independent of M-II was not unconnected to the huge amount offunds approved by CECD for the BCDA, which in the minds of some M-II staff ought to be managed directly by M-II instead of allowing BCDA to manage the funds to run the Association's engagement in the process.

Gradually however, what started as partnership arrangement manifested to something else that requires development experts, organizational management experts and corporate leadership experts to conduct further research to determine whether the brand of partnership propelled by M-II can actually manifest to true partnership arrangement to ensure sustainable extractive sector management.

M-II which came to the Basil Community leaders with idea of partnership in the implementation of CECD Corporate Social Responsibility initiative has made a 360 degree turn and has taken over control of the partnership arrangement by "starving BCDA of funds". The BCDA is now run by M-II with their Policies not BCDA Policies.

Issue No. 2: Draft 2013 Work Plan and Budget

During the December 20, 2012 Basil Community Town hall meeting when the leadership of BCDA was inaugurated, the BCDA was asked to submit their 2013 annual work plan and budget to CECD Board of Directors for consideration and approval.

When the BCDA drafted the 2013 work plan and budget as contained in the BCDA Byelaws, M-II staff requested BCDA to give them the document to flesh it out before BCDA transmits the work plan and budget to CECD Board of Directors. Instead of allowing the Chairman of BCDA whose responsibility it was to present and defend the budget before the CECD Board of Directors, M-II took the budget themselves and presented it before CECD Board of Directors.

M-II justified their reason on the fact that M-II owns BCDA and should detect what happens to BCDA, but forgetting that they sold the project to Basil Community members as partnership organization that belongs to the people Basil Community not M-II. So it could be safe to conclude that from the beginning the partnership was built on false hope, lies and deceit.

Could it be that M-II does not understand its roles and responsibilities in partnership arrangement or could this be linked to lack of managerial or leadership experience, greed, corrupt tendencies or ordinary quest for power and control? Or is it a case of betrayal of CECD efforts to ensure peaceful co-existence with their host community.

Issue No. 3: Staff Recruitment by M-II Team for BCDA

One of the conflicts between M-II and the BCDA arose from the recruitment of a staff who M-II called Basil Community Liaison Officer (BCLO) by M-II Team. The BCLO was recruited to work with and be paid by the BCDA, without informing BCDA who is mandated to facilitate, Coordinate and shape the strategic direction of Basil Community engagement in the project.

When asked by the CECD if the BCDA was aware of and participated in the recruitment of the BCLO, the Director of M-II said yes.

The reality is that BCDA was neither informed nor consulted and was not aware of the recruitment of the BCLO. If for anything, if there was any communication to that effect, the BCDA Chairman and Secretary in a transparent organization would have been the first people to know.

After BCDA staff has conducted the interview and made their selection on the candidate to be employed, they asked the BCDA Secretary to send a copy of BCDA official letter-head paper to M-II Team. At this point in time, the BCDA has not printed its official letter-head paper because M-II management refused to release funds to the BCDA to print the official letter-head paper. So there was no official letter-head paper to give to M-II Team.

The next thing the M-II Team did was to ask for electronic signature of the BCDA Chairman and BCDA Secretary. With this the BCDA Chairman and Secretary became suspicious and curious that something more ominous like fraud was brewing because as at this time the BCDA have not known anything about the BCLO recruitment.

The purpose of asking for the BCDA official letter-head paper and electronic signatures was to use the BCDA official letter-head paper to write the employment letter and copy and paste the electronic signatures of the BCDA Chairman and the Secretary respectively on the employment letter, without informing the BCDA.

However, to ensure that the recruitment process continued unhindered, the M-II Team forged BCDA letter-head paper unknowing to the BCDA which they used to write the employment letter and contract which they inserted the names of the BCDA Chairman and Secretary as the authors of the employment letter and the employment contract document.

True to the fact the employment letters and contracts were drafted and printed on the forged BCDA letter-head paper, but could not be dispatched to the recruited BCLO because the BCDA Chairman and the Secretary refused to send their electronic signatures to M-II Team.

The CECD asked the Director of M-II what was delaying the recruitment process. It was then the M-II Director informed the BCDA Chairman about the recruitment and sent a copy of his response to CECD to the BCDA Chairman whom he wanted to back him in his fabricated story should his boss eventually want to verify from the BCDA if they were aware of and participated in the recruitment exercise.

Upon receipt of the copy of the letter and reading through it and discovering the lies, deception and fraud being perpetrated by M-II Team under the leadership of the M-II Director, on the BCDA, the BCDA Chairman prepared a letter to the CECD informing the management that BCDA was unaware and did not participate in the recruitment process.

The letter from BCDA Chairman to CECD turned out to be the contextual hub through which M-II Team who supposed to facilitate the institutionalization of peaceful co-existence between CECD and Basil Community through the establishment of BCDA and transmuting it to self-reliant Peace facilitation organization in the Basil Community became "the disease they came to cure".

Issue No. 4: Campaign of Calumny against BCDA

The next step M-II Team took was to create conflict and sow a seed of discord between the BCDA Secretary and BCDA Chairman but they failed, as maturity and leadership skills have critical roles to play in organizational management. Having failed in this bid, they resorted to go into Basil Community to poison the minds of the community members, some of whom were bribed behind the scene, against their BCDA representatives, asking them to recall the BCDA members.

On the 7th June 2013, BCDA Chairman and Secretary received email from M-II Team indicating that the training billed to commence on the 13th-18th June 2013 has been cancelled.

The cancellation was not unconnected with the rift that now existed between the BCDA and the M-II Team over the recruitment of the BCLO without the knowledge of the BCDA.

On June 21st 2013, there was a reconciliation meeting to settle the above face-off between M-II Team and BCDA members at Focus Homestead Hotels in Sunbird City.

The major issue was the employment of BCLO who is to be paid salaries by the BCDA from BCDA budgetary allocations without the M-II Team informing the BCDA about the recruitment; and the same time, fabricated lies against the BCDA members.

The second issue was forging of BCDA letter-head paper and trying to fraudulently acquire and use BCDA Chairman and BCDA Secretary's electronic signature to append on the employment letter and employment contract document to be issued to the recruited BCLO. With this, the BCDA lost confidence in M-II Team as their credibility and moral compass began to point southward instead of pointing to the "True North".

At the end of this meeting, the Director of M-II Team apologized to BCDA and promised that such thing will repeat itself. He indicated that as soon as possible, the scheduled training which was earlier cancelled will take place in two weeks' time thereafter; all funds belonging to BCDA will be transferred to BCDA Bank Account.

In the month of July, M-II Team secretly invited some Basil Community members to attend a meeting, and directed the invitees to keep the proposed meeting secret. They were admonished not to inform the BCDA that represents Basil Community of the meeting and the topic of discussion was BCDA and how the BCDA Secretary allegedly forged M-II Board of Directors signature.

At that meeting the M-II Director, who started this crises by lying to CECD against the BCDA was referring to the BCDA members as "*inconsequential group of people who can be crushed and nothing will happen*".

Having been prompted, massaged and gotten their brains twisted by the M-II Team Coordinator and the M-II Director, some members of Basil Community representatives present at the meeting were literally calling for the heads of the BCDA members thereby massaging and boosting the self-esteem of the M-II Team. However, during the meeting some Basil Community members present were very concerned because they saw integrity, commitment and experience before electing their BCDA representatives to engage in the Basil Community Peace Initiative. How come now they have changed all of a sudden.

Then a question was raised as to "if the BCDA is as bad as they are projected to be, and are the main purpose of this meeting, why were they not invited to this meeting to defend themselves". The M-II Team was unable to answer the question.

At the end of the day, some delegates to the meeting believed the allegations of M-II Team against the BCDA members, while some felt indifferent to the allegations. At the end of the meeting it was decided that select Basil Community representative should hold a brief meeting in the absence of the M-II Team to actually take a critical look at the issue presented to them before they can take any action against their BCDA representatives. During this meeting, the two people who raised the issue of the forged Board Resolutions were asked to present it for everyone present to see, but none of them could present it.

At this point some people became disillusioned and called for caution because there may be more than what has been presented to them by the M-II Team. An agreement was reached that as soon as possible, the M-II Team must call for another meeting at which point the BCDA members must be invited to defend themselves of the allegations made against them by M-II Team, and if the allegations are true, the BCDA members will be recalled and removed from office.

With this, the campaign of calumny against the BCDA members especially the BCDA Chairman and Secretary became very intense all to ensure that M-II Team crumble the BCDA and ensure that BCDA cannot continue to question their lack of transparency, lack of accountability and fraudulent dispositions.

A week to the proposed second meeting, the M-II Team Coordinator requested the BCDA Secretary to send to him the copy of the Resolutions signed by the Board of Trustees, but having gotten a wind of what was being sold to Basil Community members by M-II Team, the Secretary told him that he did have any copies of the signed Resolutions and the M-II Team Coordinator believed that and never requested for the document again.

With this, M-II Team believed that since the Secretary did not have any of the signed copies of the Board Resolutions left he will not see any signed copy to tender for verification at the upcoming meeting; therefore the community representatives will believe the allegations of M-II Team against BCDA.

The second meeting was called shortly after and needless to say, the atmosphere on the arrival date was very tense. People who knew each other and have been friends from childhood, who knew each other from primary school days to University; who have collaborated with each other before in carrying out community advocacy works on other projects, could hardly speak or look at one another's face.

The campaign of calumny by M-II Team Coordinator heightened that night. Some of the hotel staff where all the participants were lodged could be heard whispering in low tones saying, "why is the M-II Team Coordinator busy going about talking about these BCDA members, destroying their names and trying to set problem for them with their community members and some of these people he is trying to rubbish are old enough to be his father. Why is he doing this to them, is it because he has more money than them or what"?

The meeting was held again at Focus Homestead Hotels in Sunbird City. In the meeting hall, the atmosphere remained very tense. From the look on the faces of some of the community members invited to the meeting by M-II Team, you could see hatred burning like fire in their eyes. Some community members present were vilifying the BCDA representatives at the meeting.

Then when the issue of the Board Resolutions was raised, the M-II Team members who were present at the meeting where the Board Chairman and Secretary signed the Resolutions denied that such issues were discussed and agreed, talk less about the drafting and signing a Board Resolutions.

At that point one could hear all sorts of comments coming from some of the community representative who have been solidly positioned by M-II Team Coordinator to lead the frontal attack on the BCDA Chairman and Secretary. This time the M-II Team could be seen "walking with their chin on their chest" with a feeling they have succeeded in discrediting and destroying the BCDA Chairman and Secretary. They were trying to literally put the minds of the BCDA members present on state of regretting being part of the Basil Community peace project.

However, the euphoria of the M-II Team and their supporters was short-lived as the BCDA Secretary reached to his folder and produced a copy of the Resolutions signed by both the Chairman and Secretary of the Board.

The next reaction from M-II Team was one of astonishment and confusion as they hurriedly grabbed the document from the Secretary to cross-check the signatures to know whether the Secretary forged the signatures as alleged; and after about 2-3 minutes of silence and confusion, they then agreed that the issue was discussed and actually the Board Chairman asked the BCDA Secretary that day to draft the Resolutions, which the Board Chairman and Secretary signed before they departed the meeting venue that day. What a shameful thing on the part of M-II Team could be heard from participants at the meeting.

At this point you could literally hear a pin drop on the floor. The room became as cold as steel and as silent as a grave. It was a massive silence wrapped with shame and soiled clothing. The M-II Team could not find a place to turn their face. It was as if they were looking for a hiding place in an empty football field. It was that kind of feeling when one has accidental bowel movement in the public. It was like someone has dig his fist so deep into their rectum and they are feeling the taste of faeces and flatus in their esophagus.

This is the point that the BCDA Chairman stood up and walked up to the M-II Director to remind him that the meeting is still on and participants are waiting in anticipation of his comments and reaction.

Though still dazed and confused and people could literally notice his unhappiness with the M-II Team Coordinator whom he had previously commissioned to retrieve every copy of Board signed Resolution from the BCDA Secretary, but was strategically out-smarted by the BCDA Secretary. At the end of what would have seemed like a very long apocalyptic day that Armageddon has refused to be postponed or end like any other normal day, the

M-II Director tried to wriggle M-II Team out of the fleabag they forced themselves into, by shamefully tendering apology whether for misrepresentation of fact and or for deception, or outright lies against the BCDA, or for wasting resources that could have been genuinely applied to do peace work in the Basil Community to heal self-inflicted injury, it did not matter. The truth has been established before the Jury invited for inquest.

After the apology has been tendered by the M-II Director to the people present at the meeting, everywhere still remained stone-cold quiet. Through this silence the voice of a community representative could be heard as he stood up and asked, "Where are those who have been commenting on what the BCDA is doing wrong and calling for their heads without hearing from them? I am yet to hear your comment. Please it is better you say something now that the truth has surfaced and lies exposed".

It was at this point that the M-II Team Coordinator tried to say something to defend his Team and their actions. The moderator tried to stop the M-II Team Coordinator from further comment on this issue since his boss has tendered apology.

The M-II Team Coordinator refused to listen to anyone and insisted he must be given audience to speak since he has not spoken since the meeting started. Then he was allowed to speak.

In his speech, he blamed the BCDA Secretary and the BCDA Chairman for causing the problems engulfing the partnership. He accused the BCDA Chairman of not being a good leader.

At this point, the BCDA Chairman who also has not addressed the floor since the beginning of the meeting was given the opportunity to present his own side of the case. He went down the memory lane, presenting all activities of the BCDA, what has been accomplished, and the challenges ahead; and the problem with the BCDA budget presented to M-II Team for review and return to BCDA for onward transmittal to CECD.

Having said all these, he proceeded to discuss the real genesis of the problems between the BCDA and M-II Team which he traced to the lies against the BCDA during the recruitment of the BCLO; how the M-II Team forged BCDA Letter Head Paper; how they used the forged letter-head paper to draft employment letter and contract for the BCLO and sort to fraudulently get hold of his electronic signature and that of BCDA Secretary to enable them write and dispatch employment letter to the recruited BCLOs. Now he has struck the nail at the head.

Following this presentation, a community representative was recognized to speak. He immediately apologized to the BCDA Chairman for his utterances against him (BCDA Chairman). He said that the M-II Team Coordinator approached him and told him how the BCDA Chairman was trying to destroy the peace initiative designed for Basil Community. He further said that the M-II Team Coordinator asked him to raise a motion for the recall of the BCDA Chairman from the BCDA and expel him from the peace project. He furthermore stated how the M-II Team Coordinator has been engaging him since he arrived in the hotel and in the restroom during the meeting and breaks to make sure he moves for the sack of the BCDA Secretary and the Chairman.

It was at this point some other attendants at the meeting started to reveal how the M-II Team Coordinator has recruited them to do the same thing.

At the end of all the hullabaloo there was an agreement that normalcy will return and the BCDA will start its training Program after which they will go back to the community and train community members.

This was also to be the last meeting held with the Technical Expert seconded by CECD who was already frustrated out by M-II and felt he had no other option than to cave in as he was being accused of supporting the BCDA with his insistence that M-II Team must follow the established protocol and assist in giving BCDA a chance to work with its Byelaws. In fact this very experienced young man could be likened to the "Manchurian Candidate of M-II Team". He was used for experiment as a guinea "pig" by M-II Team but the BCDA knew he meant well for the Basil Community peace initiative but was frustrated out by M-II Team politics.

Issue No 5: M-II Team Side-lines BCDA

The conflict between M-II Team and BCDA became so tensed up that M-II Team refused to disclose to the BCDA the amount of money approved by CECD for BCDA activities or release the funds budgeted for BCDA activities to the BCDA. Instead M-II Team went straight to implementing the activities which BCDA ought to be implementing themselves in Basil Community.
Again, during scheduled BCDA meeting in Basil Community with the Chairman of Basil Community Peace and Conflict Resolution Commission, appointed by the State Governor, M-II Team refused to provide the funds needed by BCDA to host the meeting after due authorization by M-II Board and justified this by saying "it was M-II Team's Policy not to provide funding for such meetings".

In addition, the M-II Team refused to release the money budgeted and approved for BCDA to pay for office accommodation, furnish the accommodation and recruit an Administrative staff who can handle some administrative works in the office as agreed and approved by M-II Board.

Further, the M-II Team refused to release the money that was approved for BCDA to carry out activities on the 2013 World Peace Day, instead the new M-II Team created conflicts by playing to ethnic sentiments and all these and many other unprofessional and unguarded utterances.

Issue No. 6: BCDA Quest to Reposition and Move Forward

Truth is the only gap between two parties in a disagreement. To bridge this gap requires the parties concerned to listen to each other, learn from each other and have enough thoughts, imaginations, understanding and flexibility to be able to get to the truth. This understanding prompted the BCDA as a means to refocus itself to face its mandate, to request for a meeting with BCDA Board of Trustees.

Needless to say, this meeting did not hold as the M-II Team Lead wrote to the BCDA Secretary to tell him that BCDA Byelaws does not have a provision for BCDA members to hold meeting with the BCDA Board of Trustees (BOTS), therefore there will not be any meeting with the BOTs.

Even when the M-II Team ought to have known that the two members of the Board of Trustees are also members of BCDA too, as established in Article VII. Section 7.1, of BCDA Bye Laws which states *inter alia "The BCDA shall be comprised of ten (10) elected individuals comprised of seven (7) persons representing the seven villages making up the Basil Community, two (2) representatives of the Board and one (1) representative from the M-II Team".*

But assuming these two BOT members are not members of the BCDA, is there anywhere in the world where Senior Management Team (this time the BCDA) of an organization is barred or prohibited from having a meeting with the Board of Trustees of the organization? The stance of the M-II Team in this matter speaks volume about their capacity and experience to facilitate the institutionalization of sustainable peace in Basil Community.

Now the question is, if a person is not capable of interpreting, implementing and applying the provisions of the Byelaws guiding his activities in a quest to ensure peaceful co-existence, can there be any guarantee that person can successfully lead a team that will build peace to achieve sustainable extractive sector development based on the Byelaws that could not be interpreted, implemented and applied?

The requested meeting did not hold. What the M-II Team did was to once again invite select Basil Community members to a meeting with M-II Chief Executive Officer and M-II Team to discuss a way forward for Basil Community sustainable peace project. He did not mention to them in his meeting invitation letter that the BCDA will be in attendance at the meeting.

The M-II Team also invited the BCDA members to a meeting with M-II Chief Executive Officer and M-II Team at the same venue and time, to lodge in the same hotel with the Basil Community representatives invited to the meeting without revealing this to either the Basil Community representatives or the BCDA, replicating exactly one of the tactics extractive companies used to cause conflicts in and between mining host communities.

The meeting was held on September 10, 2013 without the presence of the M-II Chief Executive Officer who was sighted walking around the meeting venue but never entered. When asked why the M-II Chief Executive Officer who requested for this meeting was not present at the meeting, the M-II Director informed the meeting attendees that they are capable of resolving any issues that might arise from the meeting, therefore whether the M-II Chief Executive Officer was there or not, the meeting should go on.

However, it is very obvious and unbelievable that the man who started the crises will now sit and superintend the resolution of the crises. So the M-II Director will listen to and resolve without bias, the complaint leveled against him and his subordinates.

Needless to say, no issues were resolved though the complaints were addressed and the only one major outcome of the meeting was the insistence by the M-II Team under the leadership of M-II Director to review the BCDA Byelaws which was adopted a year earlier which the M-II Director and the M-II Team refused to respect and implement.

The game plan of M-II Team under the overall leadership of M-II Director, who for many years served as a Community Liaison Officer for another well-known extractive company, Ox Andale Mining Company, operating in Basil Community, and who is well schooled in the act of creating intra and inter community conflicts, destabilizing and destroying host and impacted extractive communities, has always been to set the Basil Community against their elected members of BCDA; providing the enabling environment for the Basil Community and other impacted communities to engage in mutually assured destruction while M-II Team will go about unchallenged in their bid to make Basil Community sustainable peace project a failed one while they use funds allocated by CECD to fund BCDA as slush funds.

It is important that people should understand that no one can fight a war in his community and win. The simple reason is that if one fights and wins a war in his community, what about the collateral damages that follow? The people who ought to be "Peace Ambassadors" hired to help implement sustainable peace initiative should not be clamoring for war, just for selfish financial reasons or due to lack of competence to faithfully execute the project assigned to them.

But the M-II Team Director and his team brought in to facilitate sustainable peace in Basil Community simply because of money have now replaced the extractive companies in creating conflicts in the system and using the money which ought to have been spent in resolving communal conflicts and providing enabling peaceful environment for sustainable peace as panacea for economic development in Basil Community, to resolve internal organization conflict spawned by M-II Team who present themselves as peace advocates and conflict managers, thus bestowing themselves in the likes of "Conflict Entrepreneurs".

Upon returning from the meeting the BCDA Chairman resigned and he in confidence told the Secretary that his reason for resigning was that he does not want to be a party to creating more conflict in the Basil Community for his grandchildren, instead of resolving existing problems and preventing their reoccurrence.

He worked as the BCDA Chairman for 15 months without receiving a telephone recharge card. He used his personal money because M-II Team refused to release any funds to the BCDA or provide needed logistics. Prior to becoming the BCDA Chairman, he was the Secretary of Basil Community Infrastructure Development Project. His level of commitment in the project facilitated his emergence as BCDA Chairman.

The Secretary of BCDA also resigned his membership of BCDA because he did not want his integrity compromised or his years of experience rubbished as the M-II Team was becoming another source of conflict in the Basil Community and as a perceived Center for corruption and estate of fraudulent practitioners.

For the 15 months he served as the BCDA Secretary he never received any finance to buy telephone recharge card to recharge his modem and gain access to the internet or recharge to make phone calls to BCDA members; or buying printing paper or ink cartridge. Several requests sent to M-II Team for fund to carry out her responsibilities were not honored on the mere justification by M-II Team that "she was just a volunteer".

Issue No. 7: Character Flaw of M-II Team

This partnership became one infested with lies and deception from beginning to provide a conducive environment for M-II Team to use the funds devoted for BCDA activities in the Basil Community to create conflicts instead of facilitating conflict resolution, peaceful co-existence and sustainable development while the so called volunteer partners are being setup to fight against each other.

The idea of establishing BCDA was very strategic but the people who were supposed to make it stand have refused to do so because of corruption, greed and fraudulent character. From the writing on the wall, all they were interested in was how to create conflicts and crises to subvert effective collaboration in the project while designing the strategy to convert the money meant to institutionalize sustainable peace in Basil Community into their private coffers.

Again talking about character, does anyone expect someone who worked for a major extractive company in the Basil Community; who signs MoUs with the host community on behalf of the extractive company he represented and turns around to ensures that the MoUs are never implemented; a man who has no regards for the truth and can tell lies to Almighty God while praying in the Church; a man who has no regards for human dignity; a man who has no regards for the law of the land, and a man whose moral compass points southwards instead of pointing to the "True North" to truthfully lead a peace mission that will facilitate sustainable peace in Basil Community or anywhere else in the world?

Secondly, even when the M-II Team Director was counseled about the fact that M-II Team and BCDA are two different corporate bodies and should be allowed to function thus without one entity maligning and rubbishing the other in true partnership agreement, he retorted that BCDA is owned by M-II therefore as long as M-II is providing the fund for BCDA activities, M-II Team must be deciding how BCDA will function.

He however acknowledged that other funding agencies such as OXFAM, McArthur Foundation, Open Society Initiative (OSI), European Union, USAID, USADF, German EED etc. do not directly implement their partners' projects or directly manage the day-to-day affairs of their partners; it does not mean M-II will tow the same line. The decision of M-II Team to review the BCDA Byelaws was not unconnected with this line of thinking.

M-II Team was created to interface and facilitate the transition of the management of BCDA as contained in the "Concept Note" approved by CECD to the BCDA upon the approval of the BCDA Byelaws and inauguration of the BCDA.

Instead of handing over to BCDA and creating a space for peace in the Basil Community, M-II Team went around promoting strife because they did not want to handover to BCDA, simply because of selfish reasons.

Issue No. 8: M-II Team Claims BCDA Lacks Trust/Capacity

M-II Director said openly that he could not trust the BCDA to manage the funds approved for BCDA activities and that the BCDA lacks the capacity to manage the funds and even produce policy manuals and strategic plan for the BCDA.

The problem with this lack of trust for BCDA started at the point when this M-II Director himself lied to his by telling him in his response to the query issued to the M-II Director during the recruitment of the BCLO that the BCDA was aware of every step in the recruitment process and gave its approval. This was when the Chairman of the BCDA refused to join the M-II Director in his lies and decided to tell the truth on the matter.

On the question of lacking capacity to manage funds, facilitate, coordinate and shape the strategic direction of the BCDA, it suffice to say here that all the constituent members of the BCDA were experienced that is why Basil Community elected them in the first place.

All the members of the BCDA have served in various capacities in various developmental projects in Basil Community before being elected by the community to represent them as members of the BCDA.

They have excelled in their various professions as fish farmers, retired Manager of an Investment Bank, member of state Commission on Ethics, Systems Monitoring, and Management Review; served as a Senior Policy Analyst; Private Secretary to a state governor, Senior Assistant to a Governor on Youths Affairs among others.

These are the quality of persons the M-II Director said cannot be trusted and lacks the capacity to facilitate, coordinate and shape the strategic direction of BCDA.

Issue No. 9: BCDA STRATEGIC PLAN & POLICIES

"There is the need to start the process of formulating a 5-Year strategic plan for Basil Community Development Association (BCDA) in the Basil Community".

Vividly, this was said by the Chairman of M-II Board of Trustees at the emergency BCDA meeting with the BOTs. He asked the BCDA members to develop a 5-year strategic plan for BCDA and submit the document for ratification.

Immediately thereafter, BCDA designed a work plan to start the process of developing the five-year strategic plan. This plan was forwarded to the M-II Team, but it was turned down for reasons best known to BCDA.

Instead the M-II Team decided to develop the five-year strategic plan themselves using one and a half day to provide training to Basil Community members on Entrepreneurial Leadership skills and Strategic Planning. Is this the type of capacity required to strengthen the capacity of Basil Community?

Issue No. 10: M-II Team Claim Fear of Fund Mismanagement

On the fear of mismanaging the financial resources allocated to BCDA by the BCDA to carry out required intervention to achieve bilateral peace raised by the M-II Team as alluded to above when the M-II Director said he cannot trust the BCDA to manage the fund; this could be a legitimate fear in an environment where corruption and lack of capacity is endemic. But his position is baseless and is one of those spurious things to do ferment conflict whenever one needs something like power and control very desperately.

The fear entertained was after thought and may not be unrelated to self-serving compromising situation as basis of decision. The reason for this postulation was that there are four signatories to the BCDA Bank Account. Two of the signatories are "A" signatories while the remaining two signatories are "B" signatories. The four signatories comprise the M-II Director himself as signatory "A". The second signatory "A" is the Chairman of BCDA. The remaining two signatories are the BCDA Secretary and BCDA Treasurer respectively both of whom are signatories "B".

The mandate is that the two signatories "A" and at least one signatory "B" must sign the check before any money is cashed or withdrawn from the dedicated project bank account. The signed check must be sealed with BCDA official SEAL. The agreed process of operating the account and accessing the fund therein is that a letter must be generated by the BCDA Secretary instructing the Bank Manager on actions to take. This letter generated by the BCDA Secretary is sent to the BCDA Treasurer. Upon receipt of the letter, the BCDA Treasurer will prepare the check(s) as instructed in the letter. The treasurer will sign the letter and check as signatory "B". The letter and check are then sent to the BCDA Chairman and M-II Director for their signatures as signatories "A" before the signed letter of instructions to the Bank Manager and the signed check(s) are transmitted back to the BCDA Secretary to seal with the official BCDA SEAL, before transmitting to the Bank for proper action(s).

With this type of arrangement in place how on earth could the M-II Director justifiably say that he cannot trust the BCDA to manage the BCDA funds while he is a principal signatory to the BCDA account without whose signature no money will be withdrawn from the bank account and spent. If it is agreed given the benefit of doubt that the M-II Director cannot trust the BCDA, the question then is can he trust himself as a principal signatory to the dedicated bank account?

From the above information, one needs not be told that there is more than meets the eyes in the partnership arrangement between BCDA and M-II Team. This matter have been presented to some experts in peace building and conflict resolution and even some consultants that work with BCDA and some staff that have resigned from M-II and some still working with M-II in other areas, and their response is that the real issue of trust or technical capacity to deliver mooted by the M-II Director and his Team only boils down to selfishness and self-serving. It is founded on the fear of the quality of the members elected by Basil Community to serve in the BCDA because the M-II Team is afraid that their capacity is far below that of the BCDA members and therefore could lose their jobs if the BCDA is allowed to take over the management of their affairs.

Some opined that this is the classic case of calling the dog a bad name in an effort to justify its hanging. Some others believe that all the efforts of M-II Team to discredit and bring the BCDA to disrepute are a clear indication of diminished integrity and mischief on the part of M-II Team.

As a matter of fact, through the three dimensional ways (as peace advocate, as industrial/management consultant and finally as a researcher in the extractive sector) that the author of this book analyzed trickeries and watched M-II Team confusion and flip-flopping in their "approach-withdraw-management-system" unfolds, in his opinion he believes that the M-II Team has destroyed the very firm foundation that CECD laid for BCDA on a solid rock and consequently replaced it with a wobbly foundation laid on sandy beach for the BCDA edifice. In expressing his opinion again, he believes the stance of M-II Team is too insidious, too poisonous and too acidic for peace and peaceful co-existence of CECD in Basil Community, and elevates the whole CECD initiative to a higher level crisis and confrontation in Basil Community.

This type of partnership with articulated daring ventures such as repression, marginalization and cheating are chronically treacherous to peace and peaceful co-existence and sustainable extractive sector management.

Issue No. 11: M-II Team Claim Fear of BCDA Failure

Another reason the M-II Team advanced for not allowing BCDA to operate as an independent entity was the fear that a number of organizations have failed, therefore it will be very risky to allow BCDA to function as an independent organization not knowing whether they will fail or not. Yes, this may be a legitimate fear; however, they were unable to say why those other organizations failed.

In every part of the world there are quite a lot of organizations, campaigns, movements, big and small corporations, Cooperative Societies, churches, community development associations amid others that have failed. These failures can be attributed to many factors such as social, economic and political factors. Bad leadership infused with corrupt practices and lack of capacity contributed to the failure of some of these organizations. Where there is no effective operational policies and procedures, well thought out designs and implemented strategies imbued with quality commitment and integrity, there could be failure. If there is no proper guidance or facilitation, there could be failure. Where there is fear of failure and the people are not allowed to take decisions, learn from their mistakes, and correct those mistakes as they progress along there is likely to be failure.

However, the fear of the failure of BCDA was wrongly entertained because the quality of members of BCDA as stated on issue number 8 above is enough to pass a judgment that the fear of failure was not well placed.

Members of the BCDA belong to other associations that are surviving today and these associations were established and propelled to their current level of recognition with the contributions of these BCDA members. M-II and M-II Team have made some mistakes and are still making efforts to correct such mistakes. They have not failed despite their accumulated mistakes over many years and if M-II and M-II Team can stay afloat with high tidal waves of its longitudinal mistakes, why can't BCDA do the same and even better?

Conclusion

This case study is presented here because many extractive companies have been consistently fingered and are still being blamed for causing problems in the host and impacted communities where they operate their ventures, however one thing people tend to forget is that no matter how much problems these extractive companies try to create in the host communities, they can never succeed without the collusion and proclivity of the citizens and residents of the host community playing critical supportive roles.

It is also important to mention that even when the efforts of the extractive companies to become good corporate citizens are real, people from the host communities will frustrate such effort as perceived from this case study with the attempt of M-II Team to destroy and rubbish the BCDA in the Basil Community.

Recommendations

Firstly therefore, unless M-II Team stops thinking of how to survive and provides enhanced opportunity to the BCDA and M-II Team to learn how to lead and also understand that to facilitate means simply to "make an action or a process possible or easier" and that a facilitator "is a person who helps somebody to do something more easily by discussing problems, giving advice, or helping a process take place, rather than telling them what to do or doing it for them".

Secondly, if CECD really wants its peace initiative in Basil Community to succeed, it should rethink its strategy and advise M-II Team to divorce itself from the day-to-day management of BCDA, leaving real community people with good character, competence, capability and quality leadership/management background who really bear the daily brunt of the negative impacts of the mining project of CECD in Basil Community to manage the process of peace-building in their community.

Thirdly, send another technical expert to moderate the process, guiding, and advising both M-II Team and BCDA, for about six months at the maximum; during which time BCDA will be allowed to take their decisions, make mistakes from those decisions and learn from their mistakes, as M-II and M-II Team are still making mistakes every day and learning from those mistakes. That is a way BCDA can succeed and become institutionalized as peace advocate with some credibility Basil Community and with or without continued CECD funding.

Further, it is important CECD stops funding BCDA through M-II and channel all funds allocated to BCDA activities to the BCDA Board of Trustees who will then disburse the funds to BCDA account to be managed by BCDA as established in BCDA Byelaws. This will guarantee transparency and accountability in fund management and project implementation.

Furthermore, M-II Team should be required to give full disclosure of how the funds allocated to BCDA activities is being spent, who is spending it, and the purpose of the expenditure. This will enable CECD to understand the true picture of how the peace initiative in Basil Community is actually progressing; and the true character of those involved in the process, because there are many shenanigans playing out in this peace project in Basil Community which this book cannot sufficiently unveil.

Finally, with the unprofessional behavior and lack of capacity and integrity demonstrated by M-II Team regarding the partnership with BCDA, it is safe to say there is need to actually verify the claims of successes and achievements of the project by M-II Team and their impacts on Basil Community.

These recommendations will only be useful if M-II Team is not playing to the gallery so designed and sanctioned from the above.

But the bottom line is that:

1. M-II Team should not harm their partners for selfish reasons.
2. M-II Team should not deceive their partners for selfish reasons.
3. M-II Team should not misrepresent information to their partners.
4. There must be true partnership agreed and implemented with Basil Community on equal basis.

Printed in the United States
By Bookmasters